IMAGES of America

BUILDING MOONSHIPS
THE GRUMMAN LUNAR MODULE

A View from Lunar Orbit. This is a view from the lunar module pilot's window. The lunar module, or LM, was in lunar orbit.

IMAGES
of America

BUILDING MOONSHIPS
THE GRUMMAN LUNAR MODULE

Joshua Stoff

ARCADIA
PUBLISHING

Copyright © 2004 by Joshua Stoff
ISBN 978-0-7385-3586-9

Published by Arcadia Publishing
Charleston SC, Chicago IL, Portsmouth NH, San Francisco CA

Printed in the United States of America

Library of Congress Catalog Card Number: 2004102287

For all general information contact Arcadia Publishing at:
Telephone 843-853-2070
Fax 843-853-0044
E-mail sales@arcadiapublishing.com
For customer service and orders:
Toll-Free 1-888-313-2665

Visit us on the Internet at www.arcadiapublishing.com

For Matthew and Tyler,
who will be fortunate enough
to see the first footsteps on Mars.

Contents

Introduction		7
1.	Designing a Moonship	9
2.	Building Moonships	37
3.	Testing and Training for the Moon	69
4.	Shipping and Installation	79
5.	Missions to the Moon	89
6.	The Lunar Modules That Never Were	111
7.	Lunar Module Vehicle Logos	121

INTRODUCTION

After hundreds of thousands of years of occupancy and several thousand years of recorded history, man quite suddenly left the planet Earth in 1969 to fly to its nearest neighbor, the moon.

Although much has been written about the first men who set foot on the moon, those first hesitant steps on a world that was not our own were made possible by the efforts of designers and technicians who never left the warmth and safety of mother earth. With their own hands they built the lunar module (LM), one of the most historic machines in all of human history. The LM was a vehicle so radically new, so complex, and yet so delicate, that there were simply no rules to follow.

In 1961, after the United States had acquired a total of 15 minutes of spaceflight experience, Pres. John F. Kennedy announced his plans for landing a man on the moon by 1970. The space race had begun. In the spring of 1961, when Kennedy committed America to go to the moon, no one even knew how to get there. There were several competing plans within the National Aeronautics and Space Administration (NASA): build a giant rocket that flies directly to the moon, launch two rockets that meet in Earth orbit and assemble a lunar spacecraft that flies to the moon, or launch a spacecraft that remains in lunar orbit and sends down to the moon a specialized lander that can return and dock with the mother ship in orbit.

Through 1961 and 1962, NASA favored the direct approach to landing on the moon. As we had no experience in rendezvous and docking in space, NASA believed that this was the safest way to get there. The drawback was that a huge rocket, even larger than the 363-foot-high Saturn V, was required to boost a spacecraft that big and heavy to the moon. To develop such a rocket was going to take longer and be more costly than developing the lunar orbit rendezvous concept. In 1962, NASA decided that the lunar orbit rendezvous idea was the way to go. Although the rendezvous in orbit was a difficult maneuver far from home, the concept offered great savings in time, money, and engineering. Just one rocket was needed, with two specialized vehicles: one for the astronauts to ride in for launch and landing on Earth, and a light, simple lander for landing on the moon. This approach was championed by both NASA engineer John Houbolt and the Grumman Aircraft Corporation.

After a strenuous competition, NASA announced in 1962 that the Grumman Aircraft Corporation of Bethpage, New York, had won the contract to build the LM—the spacecraft that would take Americans to the moon—for Project Apollo. This was the first and is still the only vehicle designed to take humans from one world to another.

From the beginning, construction of the LM posed a unique set of engineering problems. It had to be a new type of vehicle performing a new type of role. Fortunately, the LM could take on any shape it needed. Unlike previous or current manned spacecraft, the LM operated only outside Earth's atmosphere, in a vacuum; it never went through any air. That is why it is such an odd-shaped spacecraft, with things sticking out all over it. The low gravity of the moon had some benefits for the LM design, too. It required less power to blast off from the moon, and it needed only a very light framework and very thin skin. By Earth standards, it seemed a flimsy contraption.

The LM was designed with several specific points in mind. It had to keep two men alive on the moon and get them there and back. It had to be small enough to fit inside the Saturn V rocket and light enough to be launched into space. It had four widely spaced legs so it would not tip over if it landed at an angle, and large footpads so it would not sink into moon dust, if there were any. The LM was a two-stage vehicle. The descent stage, with its own engine, was used to land the astronauts on the moon; the ascent stage, also with its own engine, was where the astronauts rode and lived, and was what blasted them off the moon.

At the peak of activity, in the mid-1960s, Grumman had 9,000 dedicated people working on the LM program. In Bethpage, on Long Island, each LM was meticulously assembled inside a special clean room at Grumman. This was a spotless white room (as clean as an operating room) in which the technicians building the LMs wore white suits, gloves, boots, and hats. The extreme cleanliness ensured that there would be no dirt or contamination inside the spacecraft that could short out the electronics or injure the astronauts. Unlike aircraft, the LMs were not produced on an assembly line. They were made by hand, one at a time, like fine violins. In all, each one took two and a half years to build. During the course of Project Apollo, Grumman produced 12 operational LMs. Each completed LM was carefully flown to Cape Kennedy, Florida, and most were loaded inside the giant Saturn V rocket, and launched to the moon.

On July 20, 1969, *Apollo 11*, with its lunar module *Eagle*, landed astronauts Neil Armstrong and Edwin "Buzz" Aldrin on the moon's Sea of Tranquility. In the words of Armstrong, it was "one small step for a man, one giant leap for mankind." The president's goal was accomplished. The first men on the moon landed in a spacecraft built by the hands of Long Islanders. With variations in mission objective, length of stay, cargo, and landing site, this historic event was to be repeated five more times. The last time was in December 1972. All of the LMs worked almost flawlessly, and they all got their crews home safely. During the years of intense labor—for many, seven days a week for ten years—on the LM program, the "Grummanites" had faith in themselves and in their work. This faith bordered on fanaticism. Each worker made sure the mission would not fail because of what he or she did.

In the end, Project Apollo was a bargain. It cost taxpayers a sum amounting to only one-third of one percent of the gross national product in 1970, yet the technical and scientific knowledge gained from it was immeasurable. The *Apollo* flights gave man a new sense of who he was and where he was, and the views of Earth from space dramatically portrayed the planet's fragility for the first time. July 20, 1969, was truly "one priceless moment in the history of mankind."

All photographs, unless otherwise noted, are from the collection of the Cradle of Aviation Museum.

One
DESIGNING A MOONSHIP

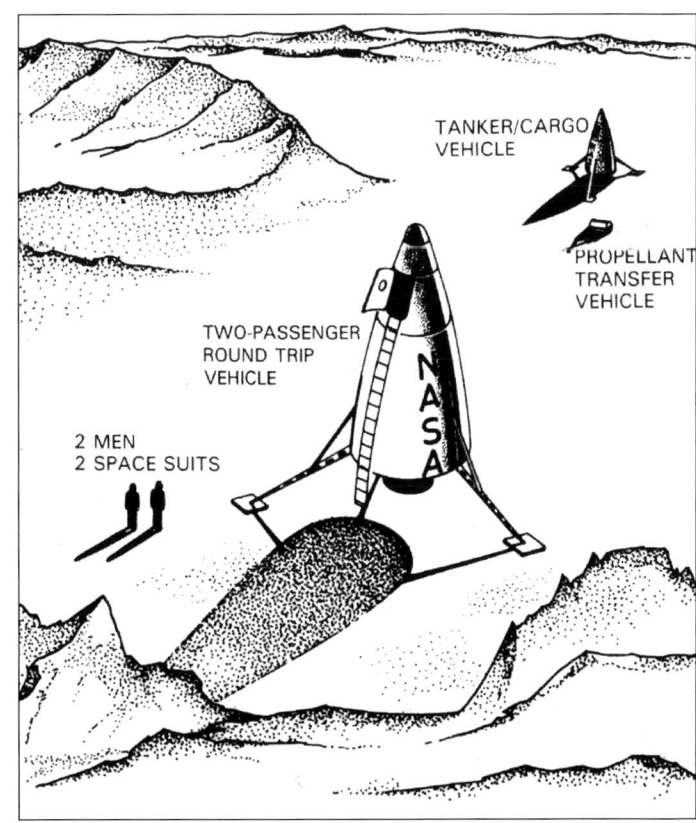

A NASA MOONSHIP CONCEPT, 1960. This early NASA concept features a moon lander that has flown directly to the moon, where it must land next to a tanker in order to refuel for the trip home.

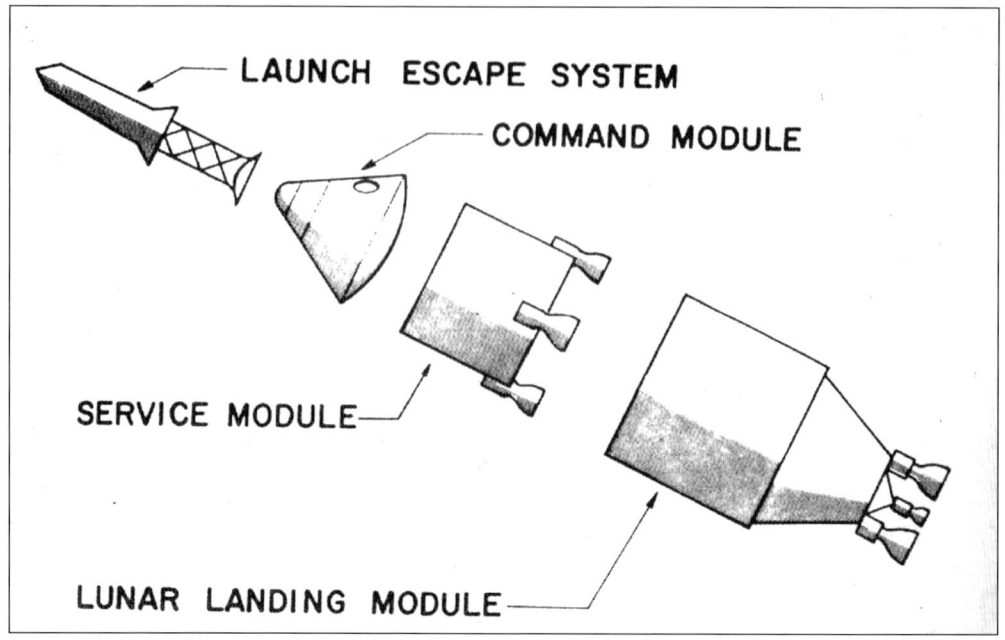

A NASA Moonship Concept, 1961. This NASA direct landing concept would have required the heavy *Apollo* command module (CM) to land on the moon, with both a descent stage and an ascent stage below it.

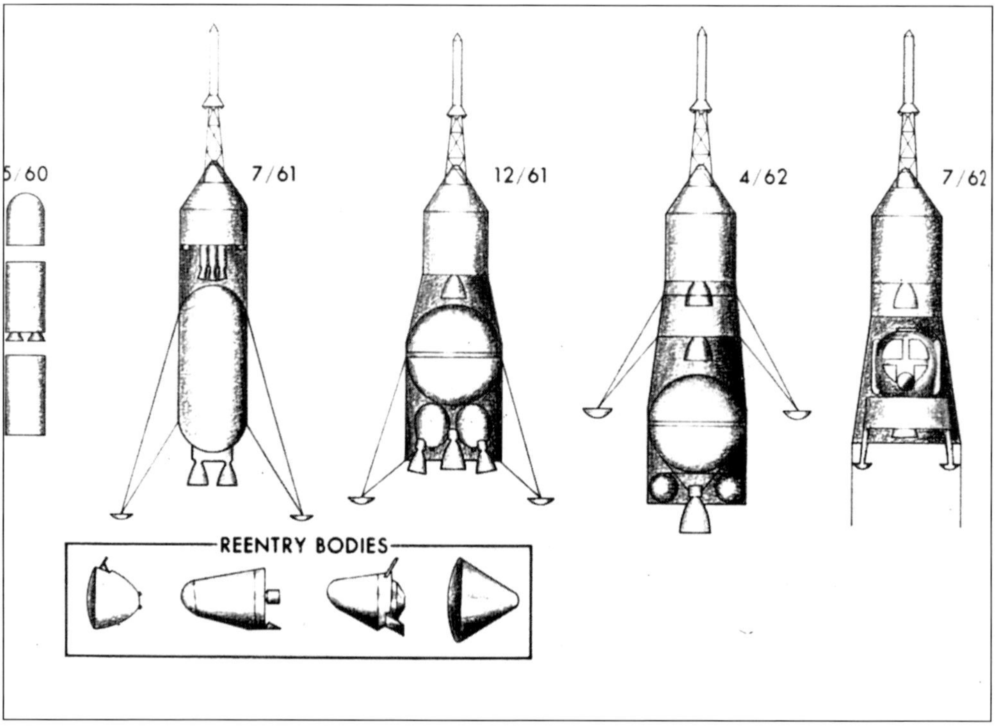

The NASA Moonship Concept Evolves, 1961–1962. From left to right are the steps by which the moonship concept evolved: from a single vehicle that went from the earth to the moon, to two separate, specialized vehicles, one of which never lands on the moon.

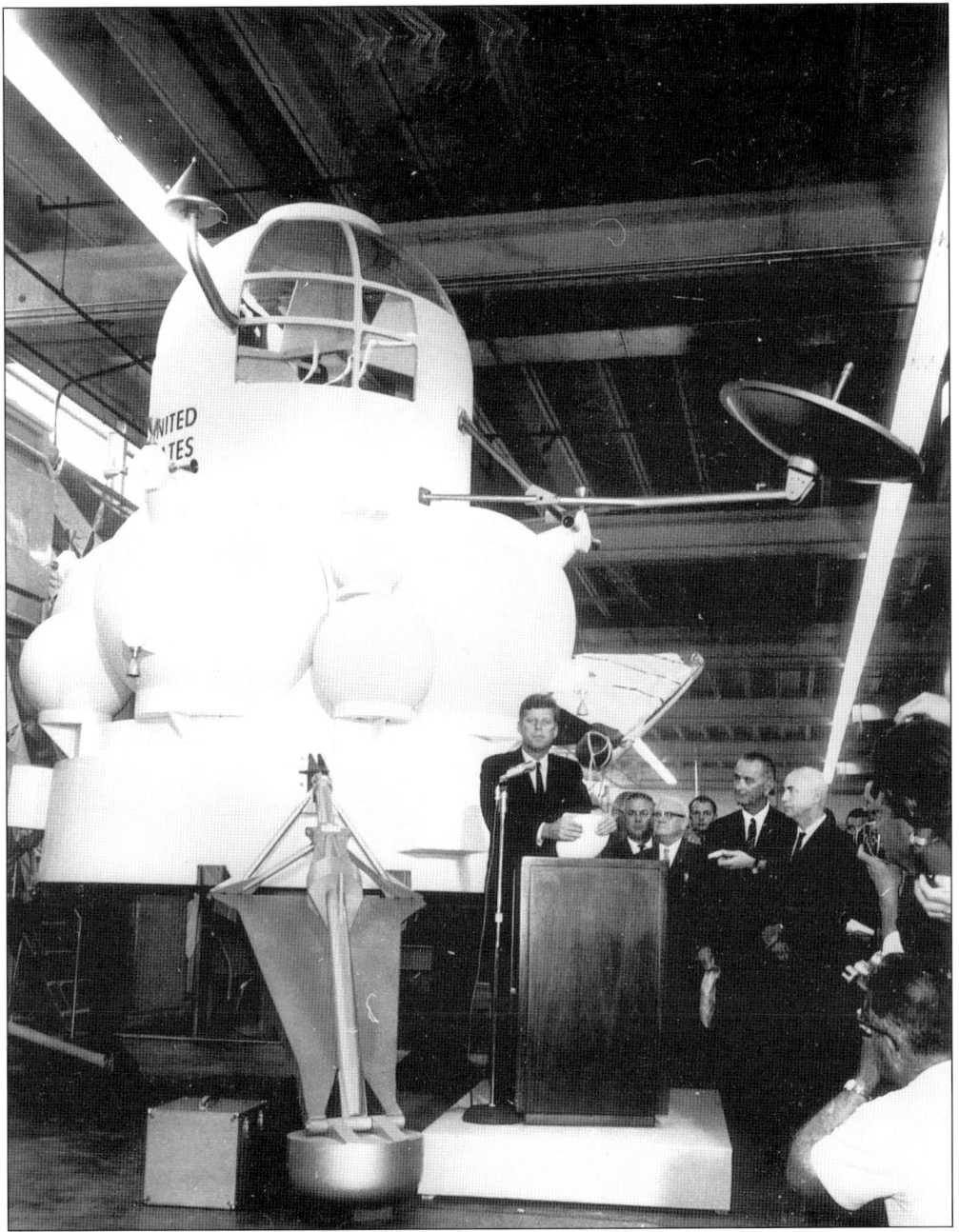

PRES. JOHN F. KENNEDY IN FRONT OF THE NASA LUNAR LANDER CONCEPT. This was NASA's first try at the design of a specialized lunar landing vehicle. Through the summer of 1962, eight aerospace companies fought to win the lunar landing vehicle contract. NASA's concept, seen here, was just a baseline. Contractors were encouraged to come up with other designs that could fulfill the mission.

A Grumman Lunar Lander Concept, Summer 1962. Grumman's first design featured fixed landing gear, two seated astronauts, large glass windows, and two docking hatches. Its only resemblance to the final version was that it was a two-stage vehicle. It had a descent stage with its own engine, used to land on the moon, and an ascent stage with its own engine, used for returning to lunar orbit. Using the descent stage as the launch pad made the vehicle lighter, thus reducing the amount of power needed for the return trip to orbit.

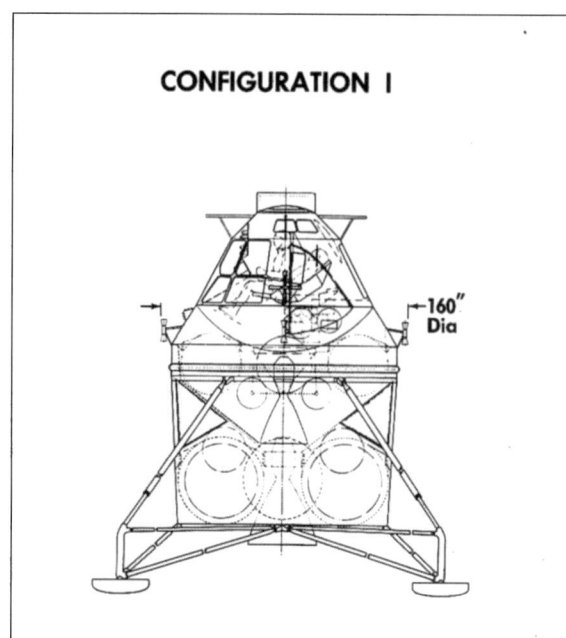

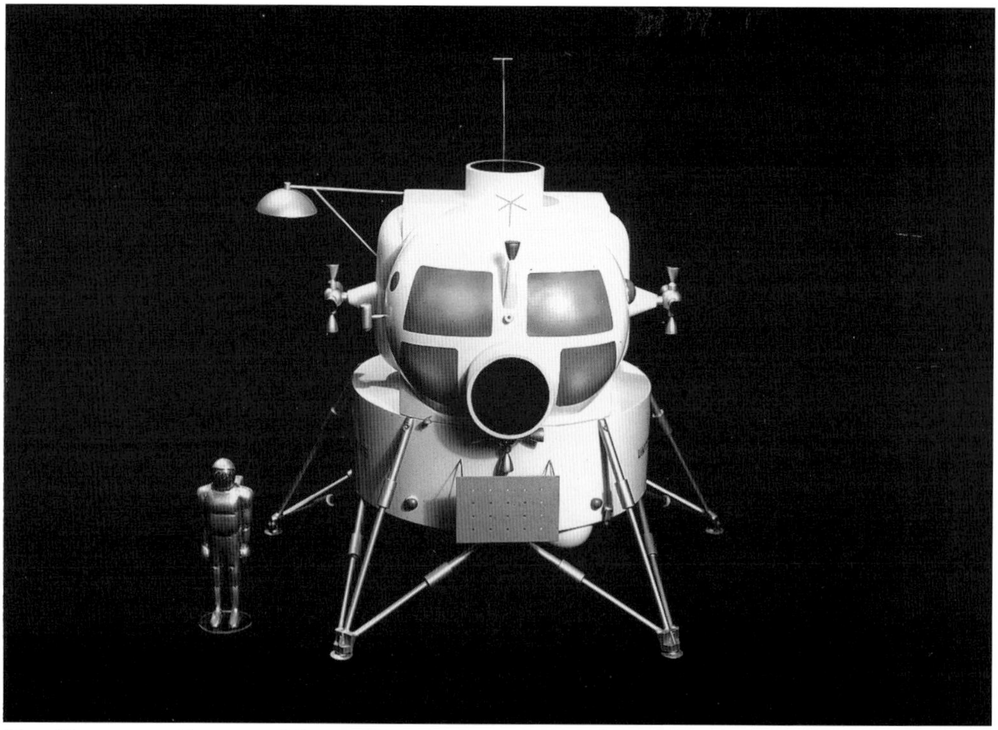

The Grumman Concept Model, Fall 1962. This small model was Grumman's final concept of a lunar lander for the contract proposal. The spacecraft featured five fixed landing legs, two round docking hatches, large windows resembling helicopter windows, and no ladder. Based on the thoroughness and brilliance of Grumman's engineering, in November 1962, NASA awarded the company a $1.61 billion contract to build 15 flight vehicles, 10 test vehicles, and 2 simulators. The race was on.

LEROY R. GRUMMAN, COMPANY FOUNDER AND CHIEF EXECUTIVE OFFICER. The Grumman Company of Bethpage, New York, had earned a solid reputation, primarily through being a supplier of naval aircraft. Grumman's fighters were among the best of World War II. The company continued building front-line naval jet fighters through the 1950s. Although Grumman was new to the space program in 1962, so was everyone else. Leroy Grumman guided the company from building biplanes in the early 1930s to the development of the LM in the 1960s.

TOM KELLY, "THE FATHER OF THE LUNAR MODULE," C. 1966. Tom Kelly was Grumman's design genius. He oversaw the LM from the initial concept through the successful lunar landing. His LM design has since been referred to as "elegant engineering." It was not pretty, but it was simple, strong, and purely functional.

TOM KELLY WITH NASA OFFICIALS, 1964. Tom Kelly (second from the left) explains the complexities of the folding LM landing gear. Listening are, from left to right, Dr. Joseph Shea (NASA), Bob Gilruth (NASA), and Joe Gavin (Grumman vice president, LM program).

A Program Management Meeting with NASA Officials. From left to right are Saul Ferdman, Joe Gavin (Grumman vice president, LM program), Eberhart Rees (NASA), Wernher Von Braun (NASA), and Tom Kelly. In this meeting, which took place on October 6, 1964, Grumman presented the first full-scale LM mock-up to NASA and went over its details. NASA had to approve of the design every step of the way.

A GRUMMAN RIGEL MISSILE, C. 1952. Grumman's first experience with rockets was on the navy's Rigel program in the 1950s. This was one of the earliest American ramjet missiles; it led to the development of later submarine-launched missiles.

A NASA ECHO SATELLITE TEST, C. 1960. Grumman's first involvement with NASA came when the company won the contract to build the launch adapter and canister for the *Echo* satellite. *Echo* was NASA's first communications satellite. Deployed from its canister and inflated in space, it was a passive relay that bounced signals.

A NASA-Grumman OAO Satellite, c. 1966. Grumman's first spacecraft was the Orbiting Astronomical Observatory, which was NASA's first space telescope. Two of these telescopes operated for many years, providing the best deep-space photographs to date. This program provided Grumman with its first clean room construction experience.

A GRUMMAN LUNAR LANDING CONCEPT, 1962. When lunar landings were first envisioned in the early 1960s, officials thought that a moon base would be set up rather than just landing one vehicle in one place for a short time. Thus, Grumman planned to make the lunar lander (upper right) a flexible vehicle by allowing it to ferry a laboratory and shelter (center) and a large pressurized mobile laboratory (lower right).

A GRUMMAN MOLAB CONCEPT, 1962. Grumman envisioned a large pressurized mobile laboratory (Molab) operating in conjunction with its lunar lander. This would allow the astronauts to live on the moon for weeks and explore a considerable area.

A GRUMMAN MOLAB PROTOTYPE, 1964. NASA funded the development of one full-scale Molab prototype, which gathered some important information on the handling of vehicles on a simulated lunar surface at Grumman's Calverton, New York, plant. Although the economics did not allow for the development of such a vehicle, the highly original and successful wheel design was retained by NASA and was ultimately used on the Mars exploration rovers in 2004.

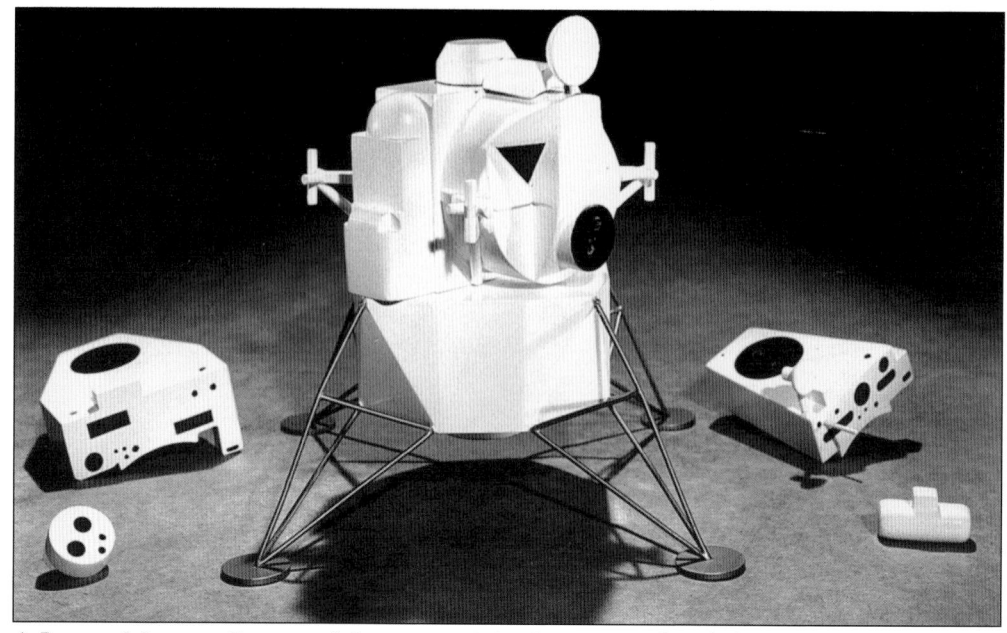

A LUNAR MODULE CONCEPT MODEL, C. 1963. It was soon found that weight was extremely critical on the LM. The heavier it was, the more fuel it had to burn while looking for a safe place to land, and the fuel capacity was very limited. Thus, one of the first things Grumman did to save weight was to eliminate the large heavy windows and substitute small triangular ones. If the astronauts kept their faces close to the small windows, they would have the same field of vision.

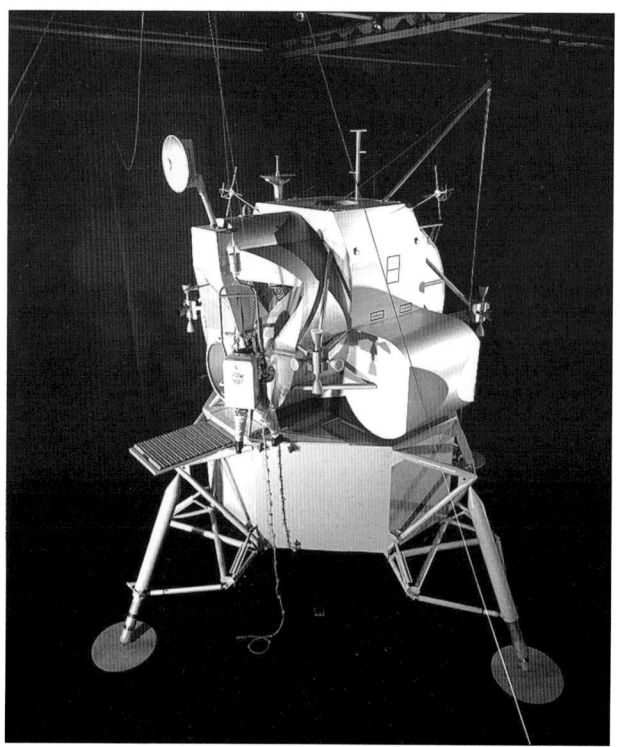

THE TM-1 MOCK-UP, MARCH 1964. The LM's design was now greatly refined. It was still a two-stage vehicle, with the descent stage serving as the launch pad for the ascent stage. In 1964, Grumman unveiled its wooden TM-1 mock-up to NASA. In order to save weight, the LM now had four folding legs instead of five fixed ones. The LM still lacked a ladder, as it was felt that if it bent upon landing, there would be no way for the crew to descend; thus, a knotted rope was supplied instead. Astronauts, however, found it impossible to climb up the knotted rope.

THE TM-1 MOCK-UP REVIEW, APRIL 1964. Due to accessibility concerns, the LM now has a ladder down the side, but it still has the round front docking hatch. As the landing legs had to be able to absorb great impact upon landing, they were fitted with internal crushable aluminum honeycomb cartridges. This was much lighter and simpler than a hydraulic system, and it only had to work once. One of these cartridges is shown being held.

THE M-5 MOCK-UP, OCTOBER 1964. In this mock-up, the first all-metal one, the LM nears its final form. The vehicle now has a ladder down the front leg, and the size of the footpads has been enlarged, as there was great concern about the hardness of the lunar surface. As the LM was going to fly only in the vacuum of space, it did not need to have any aerodynamic qualities whatsoever. Thus antennas, legs, rockets, and supplies could be attached externally wherever they were needed.

ASTRONAUTS TRAINING ON THE M-5 MOCK-UP, OCTOBER 1964. Astronauts participated in every step of the LM's design, and they were instrumental in encouraging many changes. M-5 was the product of two years of configuration studies. Every part of the LM's system was scrutinized. One of the astronauts' major concerns was that they had great difficulty getting their square backpacks in and out of the round hatch.

WERNHER VON BRAUN EXAMINING THE M-5 MOCK-UP, NOVEMBER 1964. NASA was extremely pleased with the M-5 and in the end suggested only 120 minor changes. The design had to be frozen shortly so that actual construction could begin in 1965. The crew cabins' two small triangular windows are plainly visible. Eliminating the large glass windows saved weight and simplified construction. The crew members just had to keep their faces close to the small windows to have the same field of vision the larger windows offered.

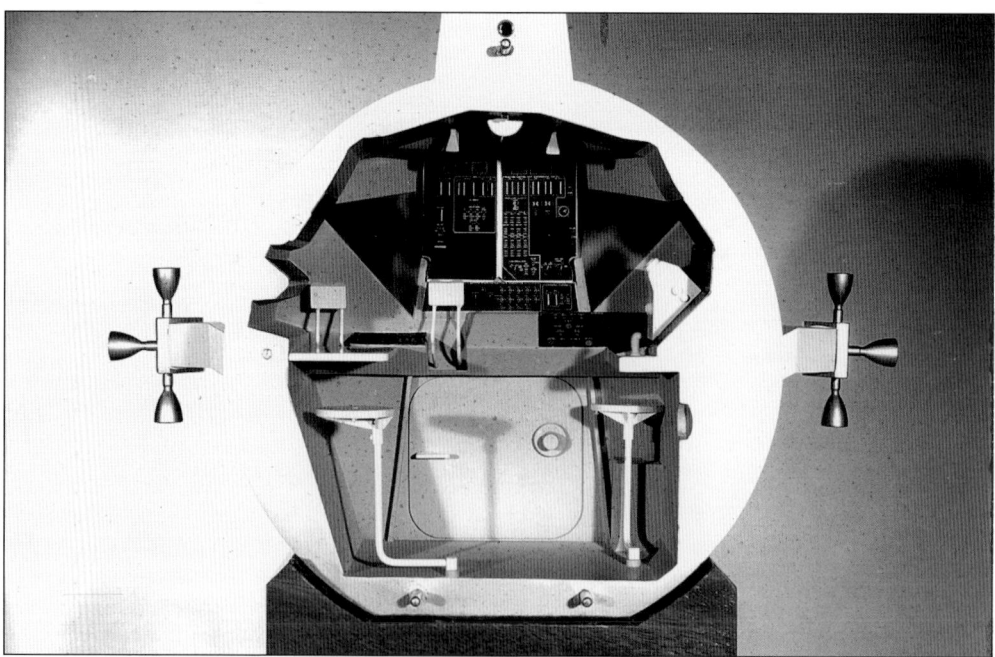

A LUNAR MODULE COCKPIT MODEL, 1964. The LM's designers originally conceived of the cockpit as having seats, like any other spacecraft. However, these were discarded, both to save weight and because they were not really needed. During landing the astronauts stood with their faces near the windows for better visibility. Grumman engineers devised a restraint system to hold the crew members in place during the landing when they were in zero gravity. Using this system, they were automatically in the proper position when they picked up the moon's gravity.

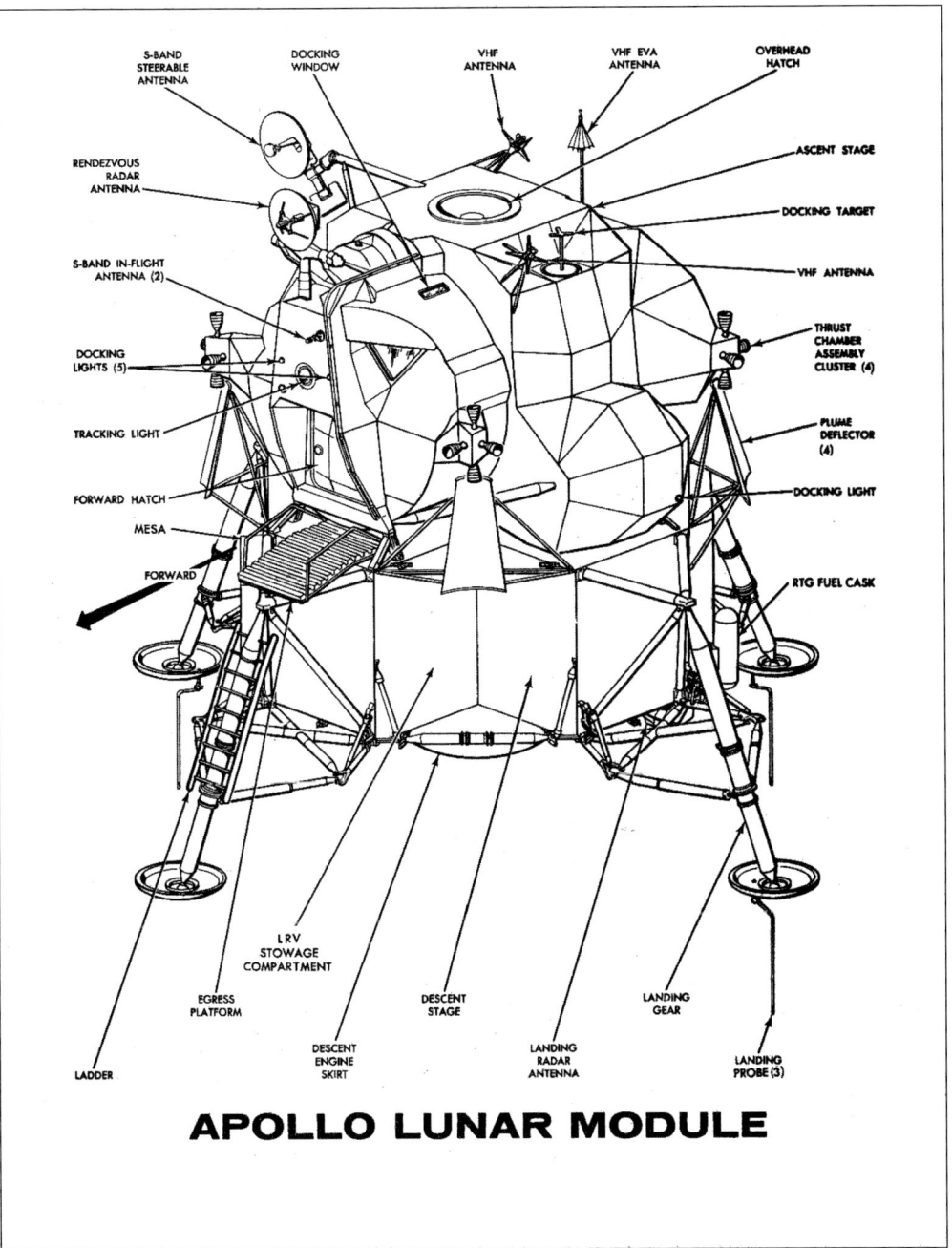

APOLLO LUNAR MODULE

THE FINAL DESIGN OF THE LUNAR MODULE, 1965. The LM design was now frozen, and construction of test and flight spacecraft could begin. The basic structure of the descent stage was cruciform, with four propellant tanks; the ascent stage was cylindrical, with a flat face and two propellant tanks. The vehicle had 18 rocket engines, the ascent stage, the descent stage, and 16 reaction-control thrusters for maneuvering.

THE FULL-SCALE LUNAR MODULE MOCK-UP, CALVERTON, NEW YORK, 1965. Astronauts train on the mock-up at Grumman's simulated lunar surface on Long Island. The LM now has a square front hatch, which greatly eased access and eliminated the heavy front docking hatch. The asymmetrical appearance of the ascent stage is due to the fact that the fuel and the oxidizer had different weights. In order to keep the spacecraft's thrust along its center of gravity, the heavier oxidizer tank (left) had to be kept in tight, while the fuel (hydrazine) tank had to be moved outboard.

AN ARTIST'S CONCEPT OF THE LUNAR MODULE IN LUNAR ORBIT, 1965. Here is how the LM in action was envisioned by Grumman. It is now in lunar orbit, its legs deployed, with two astronauts inside it and the third inside the docked CM.

AN ARTIST'S CONCEPT OF THE LUNAR MODULE ON THE MOON, 1965. The LM has now landed on the jagged mountains of the moon (there were none). Its footpads have disappeared into the lunar dust (there was very little of that). An astronaut has descended the ladder and is collecting rock samples.

AN ARTIST'S CONCEPT OF LIFTOFF FROM THE MOON, 1965. Its time on the surface now over, the LM's ascent stage blasts off, using the descent stage as its launch pad.

AN ARTIST'S CONCEPT OF THE LUNAR ORBIT RENDEZVOUS, 1965. The ascent stage is now about to dock in orbit with the CM. The commander on the left side of the cockpit can look straight up through a small window to check on alignment with the CM for docking. After the astronauts with their lunar samples transfer back to the CM, the unmanned ascent stage will be crashed back into the moon as part of a seismic experiment.

SPECIFICATIONS OF THE APOLLO SPACECRAFT. The LM stands 22 feet 11 inches tall and 31 feet across its landing legs. It weighs but 10,800 pounds when empty and 36,100 pounds when fully fueled. This heavy volume of fuel was needed to slow the LM for landing, allowing it to maneuver to a safe landing site. It was also needed for liftoff from the moon. The weight of the fuel is the reason that the vehicle had to be as light as possible.

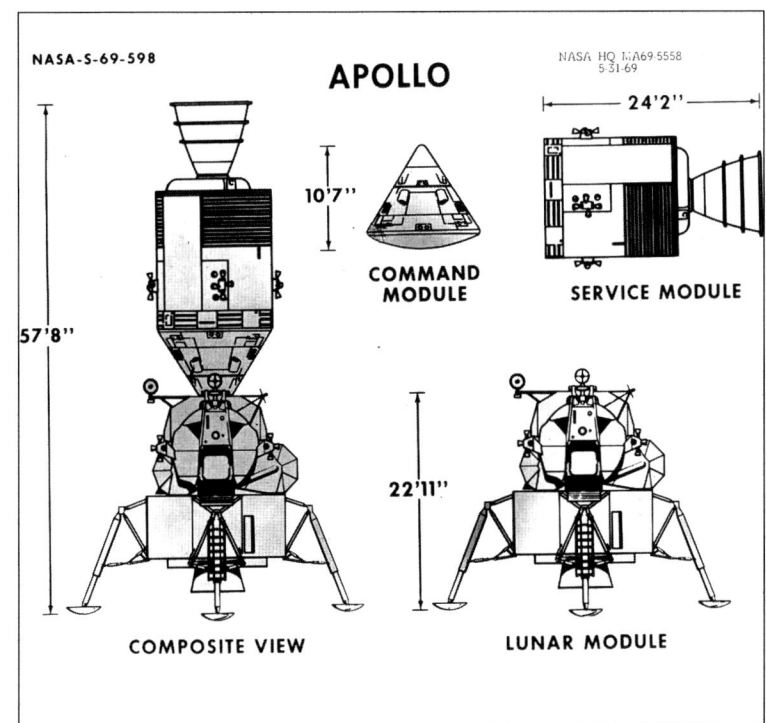

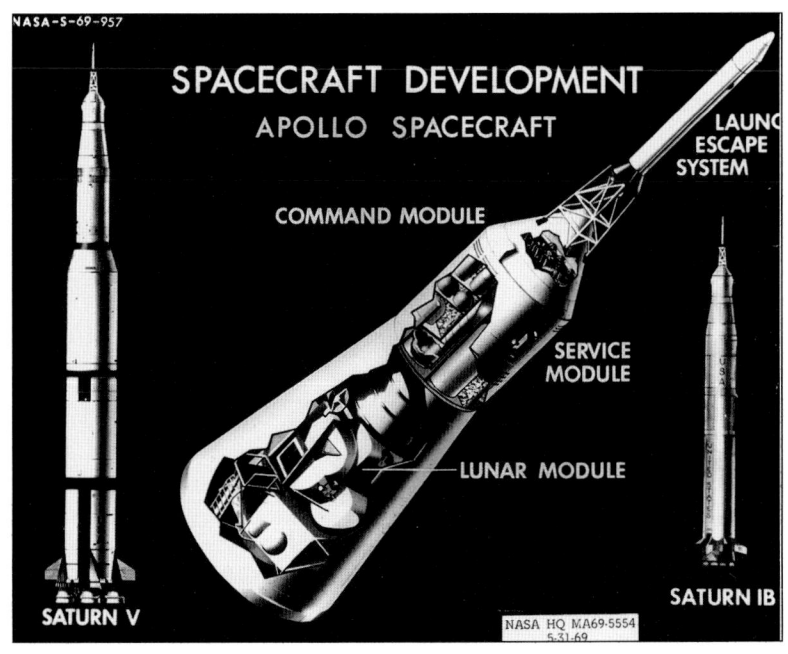

THE LUNAR MODULE INSIDE THE SATURN V ROCKET. The LM rode inside the spacecraft LM adapter (SLA) atop the third stage of the Saturn V. On the way to the moon, the astronauts would rotate the CM and pull it free.

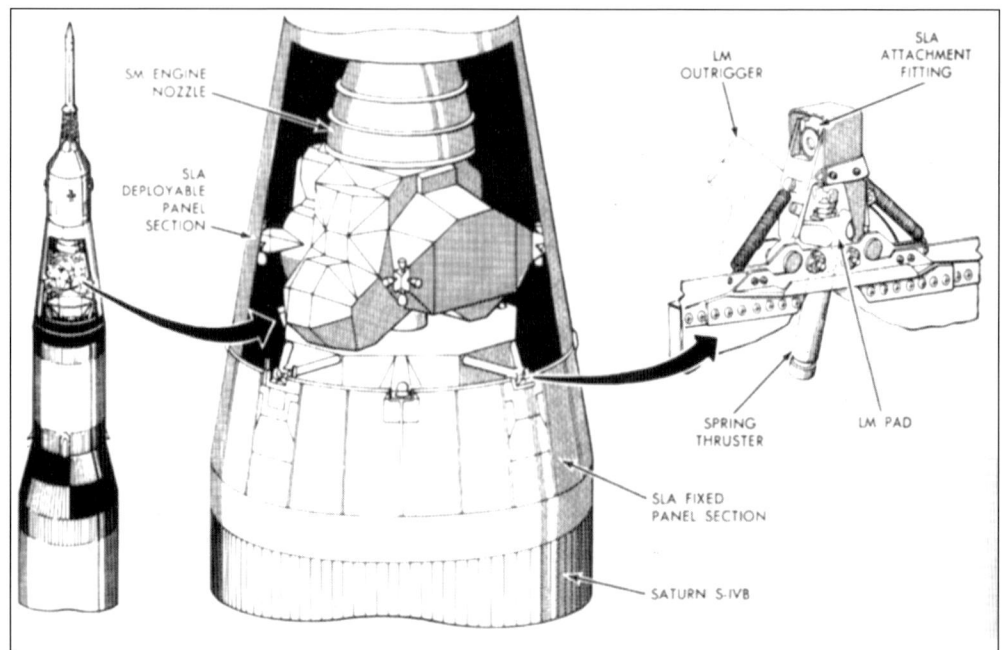

The Attachment of the Lunar Module inside the Spacecraft Lunar Module Adapter. The LM was mounted at four attach points at the top of its main landing gear struts. Explosive bolts kept it in place until the CM docked with it and it could be extracted. Once it was extracted, the spring-loaded landing gear could be deployed.

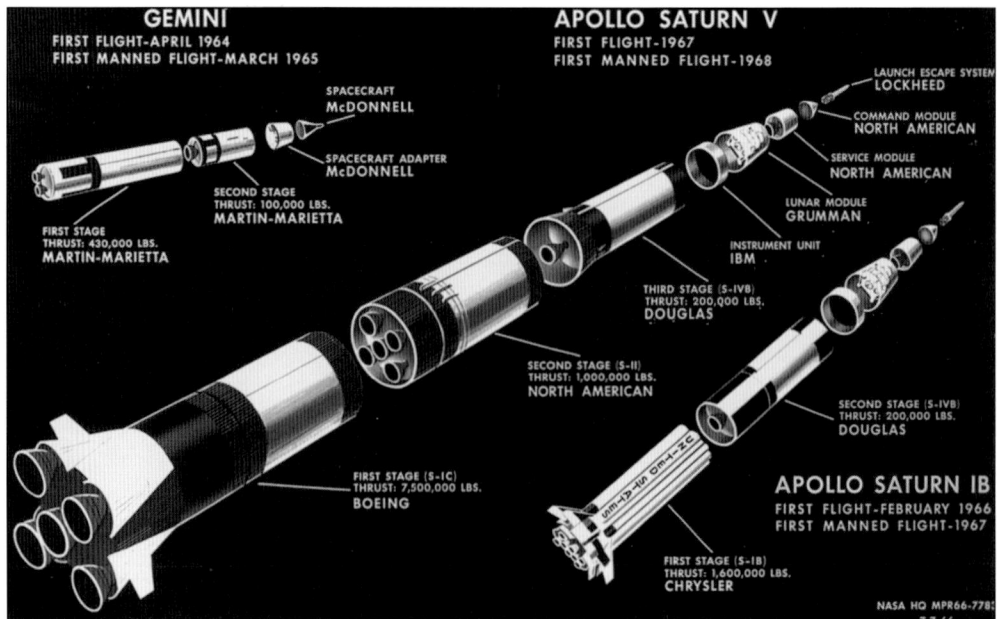

Saturn V Manufacturers and Stages. The Saturn V is still the most powerful rocket ever built; it stood 363 feet tall and had 7.5 million pounds of thrust. The rocket was such a large, complex vehicle that each major component had one prime contractor with hundreds of subcontractors. It required two stages to reach Earth orbit and a third to send it on its way to the moon.

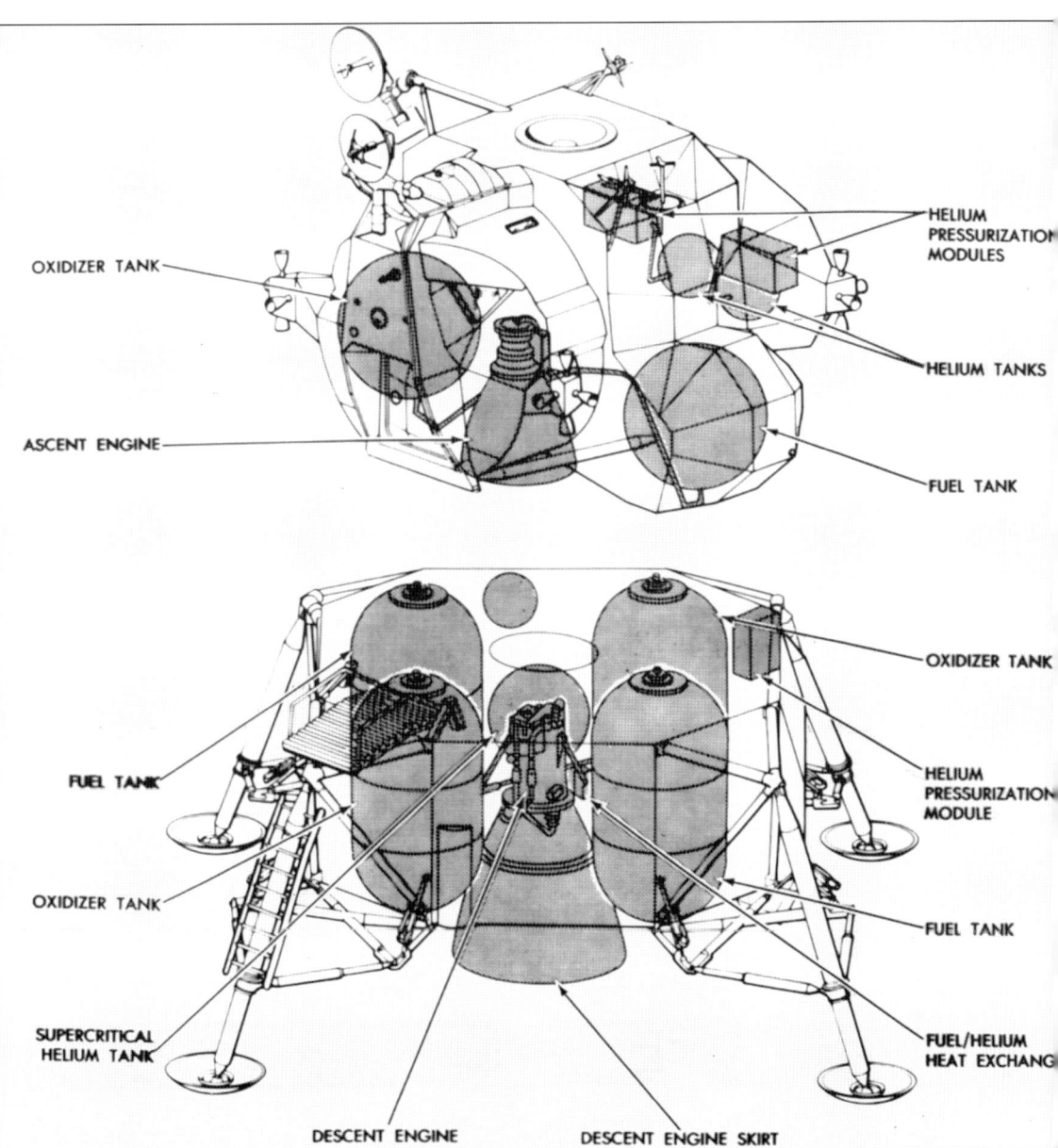

The Location of the Main Propulsion Components inside the Lunar Module. The large descent engine, built by TRW, can be seen here, with its four large tanks of fuel and oxidizer contained in the cruciform descent stage. The ascent stage has its own engine, built by Bell, with 16 reaction-control rockets, built by Marquardt. In order to keep the propulsion system simple, it was pressure-fed (it had no fuel pumps) using hypergolic propellants (which were self-igniting).

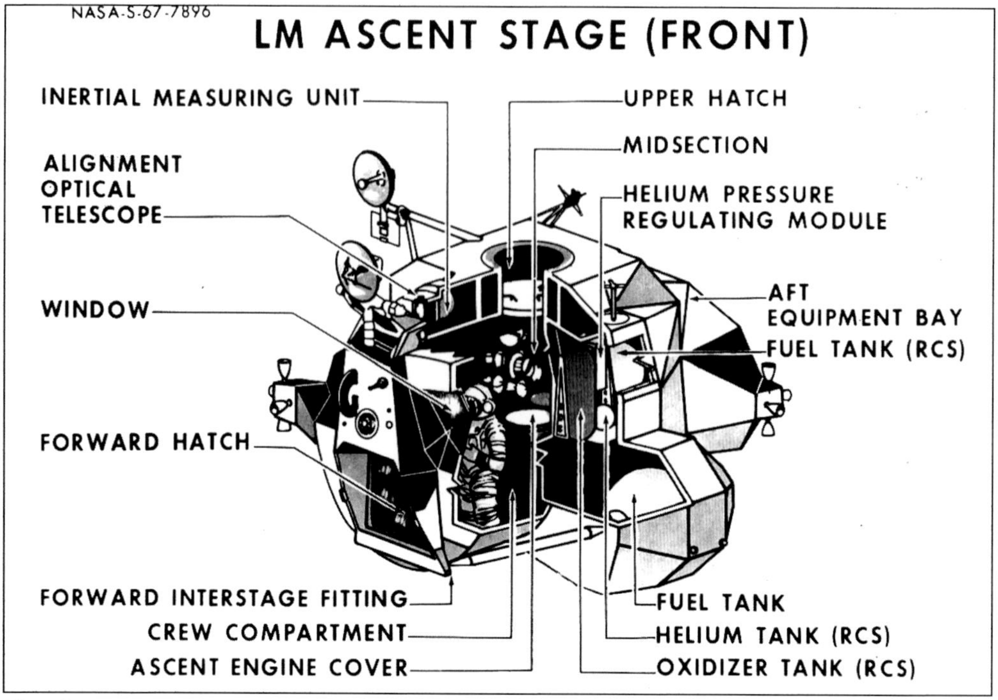

LM ASCENT STAGE (FRONT)

NASA-S-67-7896

Labels: INERTIAL MEASURING UNIT, ALIGNMENT OPTICAL TELESCOPE, WINDOW, FORWARD HATCH, FORWARD INTERSTAGE FITTING, CREW COMPARTMENT, ASCENT ENGINE COVER, UPPER HATCH, MIDSECTION, HELIUM PRESSURE REGULATING MODULE, AFT EQUIPMENT BAY, FUEL TANK (RCS), FUEL TANK, HELIUM TANK (RCS), OXIDIZER TANK (RCS)

THE LOCATION OF THE ASCENT STAGE COMPONENTS. Grumman decided on an approach of containing all of the equipment inside the ascent stage in order to provide additional micrometeoroid and thermal protection.

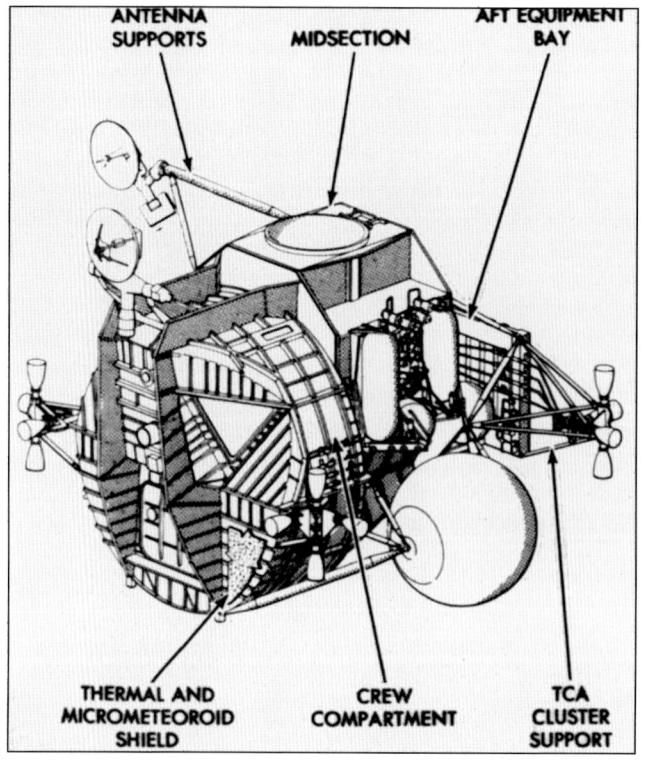

Labels: ANTENNA SUPPORTS, MIDSECTION, AFT EQUIPMENT BAY, THERMAL AND MICROMETEOROID SHIELD, CREW COMPARTMENT, TCA CLUSTER SUPPORT

THE BASIC ASCENT STAGE STRUCTURE. Grumman decided on a hybrid approach to the construction of the basic ascent stage, using aluminum alloy skins with titanium fittings and fasteners. Areas of critical structural loads were welded, but rivets were used where welding was impractical. The basic ascent stage structure was made up of three sections: the front face, the midsection, and the aft equipment bay.

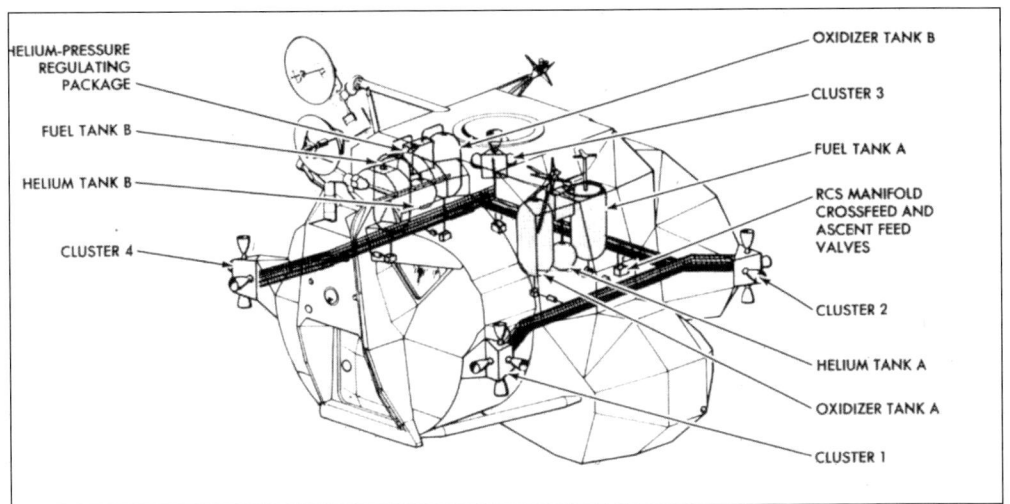

THE LOCATION OF THE REACTION CONTROL EQUIPMENT. The reaction control system (RCS) served to steer and stabilize the LM during lunar descent and ascent. The sixteen 100-pound-thrust maneuvering rockets and their dual propellant supply made up two parallel independent systems for added safety.

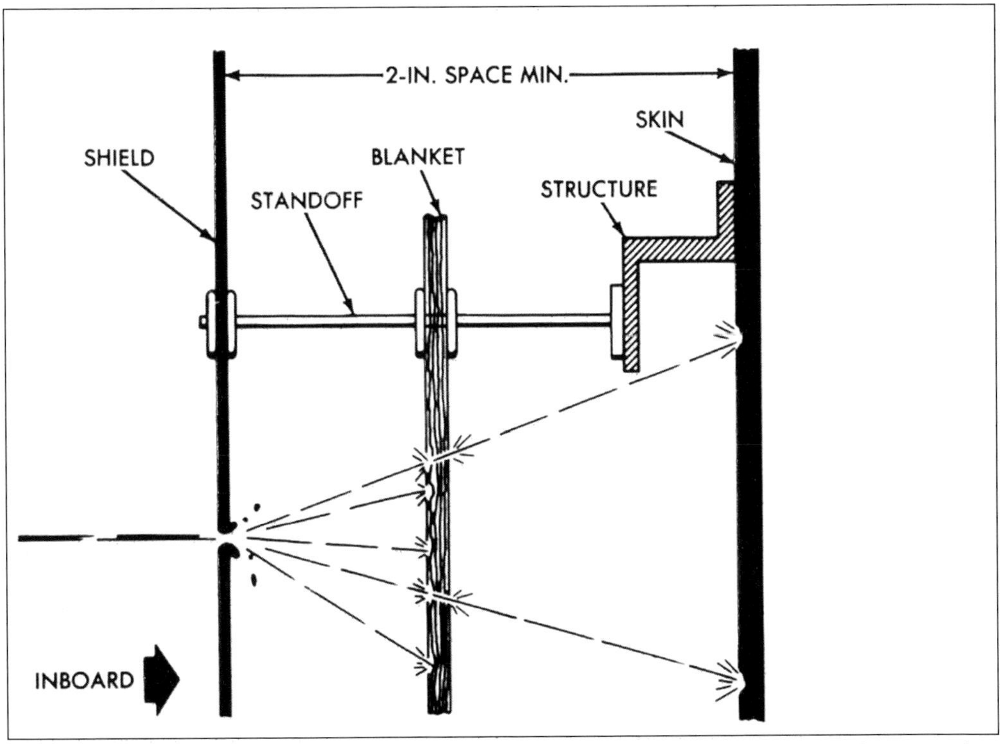

THE LUNAR MODULE'S THERMAL AND MICROMETEOROID SHIELDING. In space, the LM and its crew needed to be protected from temperature extremes (ranging up to 500 degrees Fahrenheit) and micrometeoroids. Thus, the entire ascent stage was enclosed within a thermal blanket and shielding. On the descent stage, the blanket was on the exterior. The blanket consisted of 25 layers of aluminized sheets (Mylar or H-film), which were held on standoffs from the pressurized bulkhead. Outboard of the blanket was a sheet of thin aluminum.

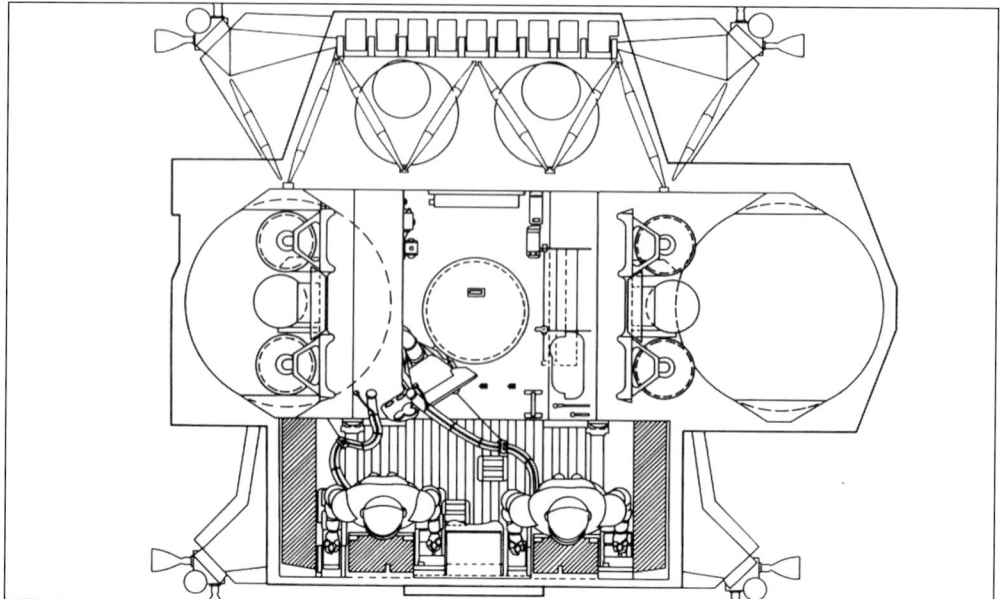

THE TOP PLAN OF THE LUNAR MODULE ASCENT STAGE. During lunar descent and ascent, the astronauts stood in the front, peering out of their respective windows. Behind them in the midsection were supplies, life-support equipment, and the ascent engine cover. In the rear were the electronics and power supply (batteries). Large spherical fuel tanks were on either side of the midsection.

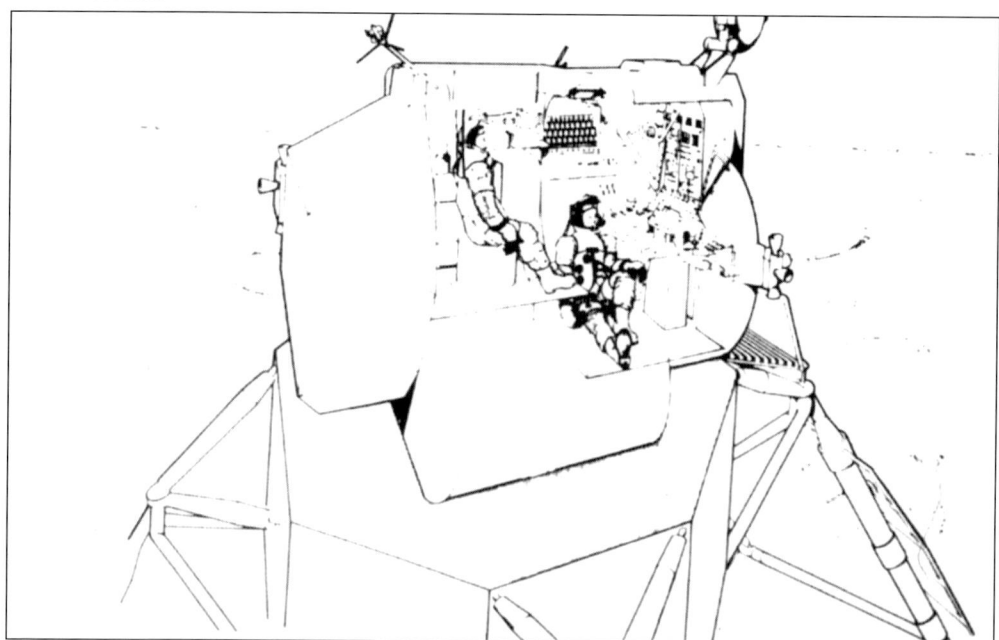

RESTING AND SLEEPING POSITIONS IN THE LUNAR MODULE. On the first missions, the astronauts rested and slept on the floor and ascent engine cover. On later missions, they had hammocks perpendicular to each other in the same general positions. While they were sleeping, outside of their spacesuits, there was only .002 inches of aluminum—about the thickness of a soda can—between them and vacuum.

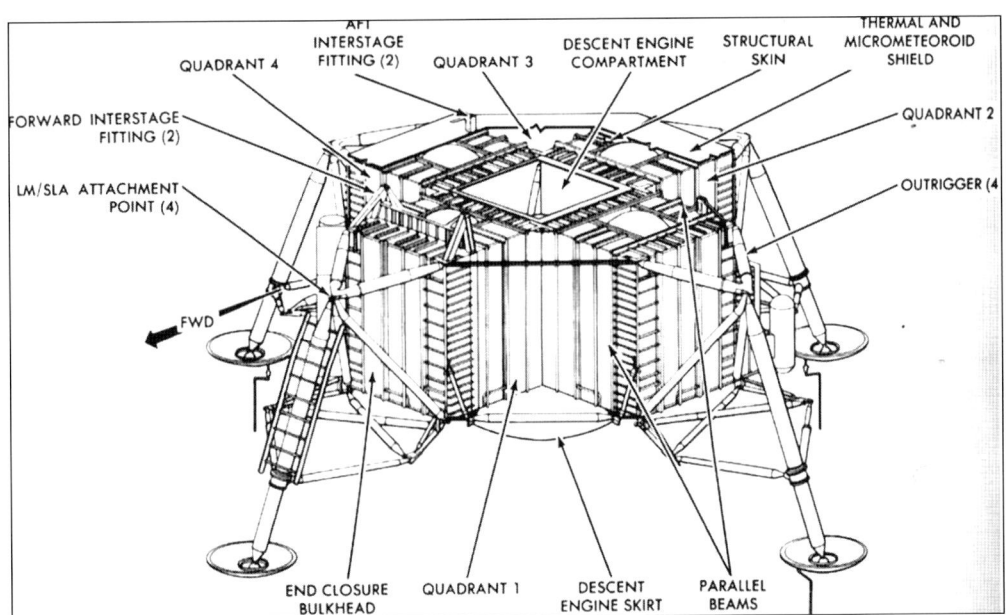

The Structure of the Lunar Module Descent Stage. The cruciform descent stage was the unmanned portion of the LM and represented two-thirds of the vehicle's weight. It had to be heavier because it supported the ascent stage; it had to absorb the impact of landing and act as the launch platform. The descent stage was covered with a heavier H-film thermal blanket than the that of the ascent stage was because it was exposed to engine exhaust from the RCS thrusters.

A Mock-up of the Lunar Roving Vehicle, Grumman, 1965. A contract for a lunar roving vehicle was also part of Project Apollo. Grumman designed the LM with the capacity to tuck a lunar roving vehicle into one of the quadrants of the LM descent stage. This photograph shows Grumman's full-scale mock-up of the lunar rover they hoped to build.

An Artist's Concept of the Deployment of the Lunar Roving Vehicle, c. 1970.
Ultimately Boeing, not Grumman, won the contract to build the lunar roving vehicle. Grumman designed the LM so it could carry the additional load externally (aerodynamics were irrelevant) and the astronauts could manually deploy it on the moon, as seen here.

Two

Building Moonships

Plant 5, Grumman's Bethpage Facility, c. 1968. This extension was built in 1964 just for the construction and testing of the LM. It contained the clean room (the dark building in the center), where three LMs could be built at once. The building in front of it contained the ACE (Automatic Checkout Equipment) Room, where the LM's testing was run. In the lower left is the Cold Flow Test Site, where the valves and tubing in the LM's propulsion and cooling systems were tested with nitrogen and glycol under pressure.

A STAND-UP MEETING, 1968. As the pressure and the schedule for building the LMs on time was extraordinary, the various managers of each vehicle met every day to go over any problems or delays that needed to be resolved. All meetings were conducted while standing so that no one would get too comfortable. From left to right are Ross Fleisig (LM-5 spacecraft manager), Al Beauregard (vehicle manager), and Ralph "Doc" Tripp (LM program director). (Photograph courtesy of Ross Fleisig.)

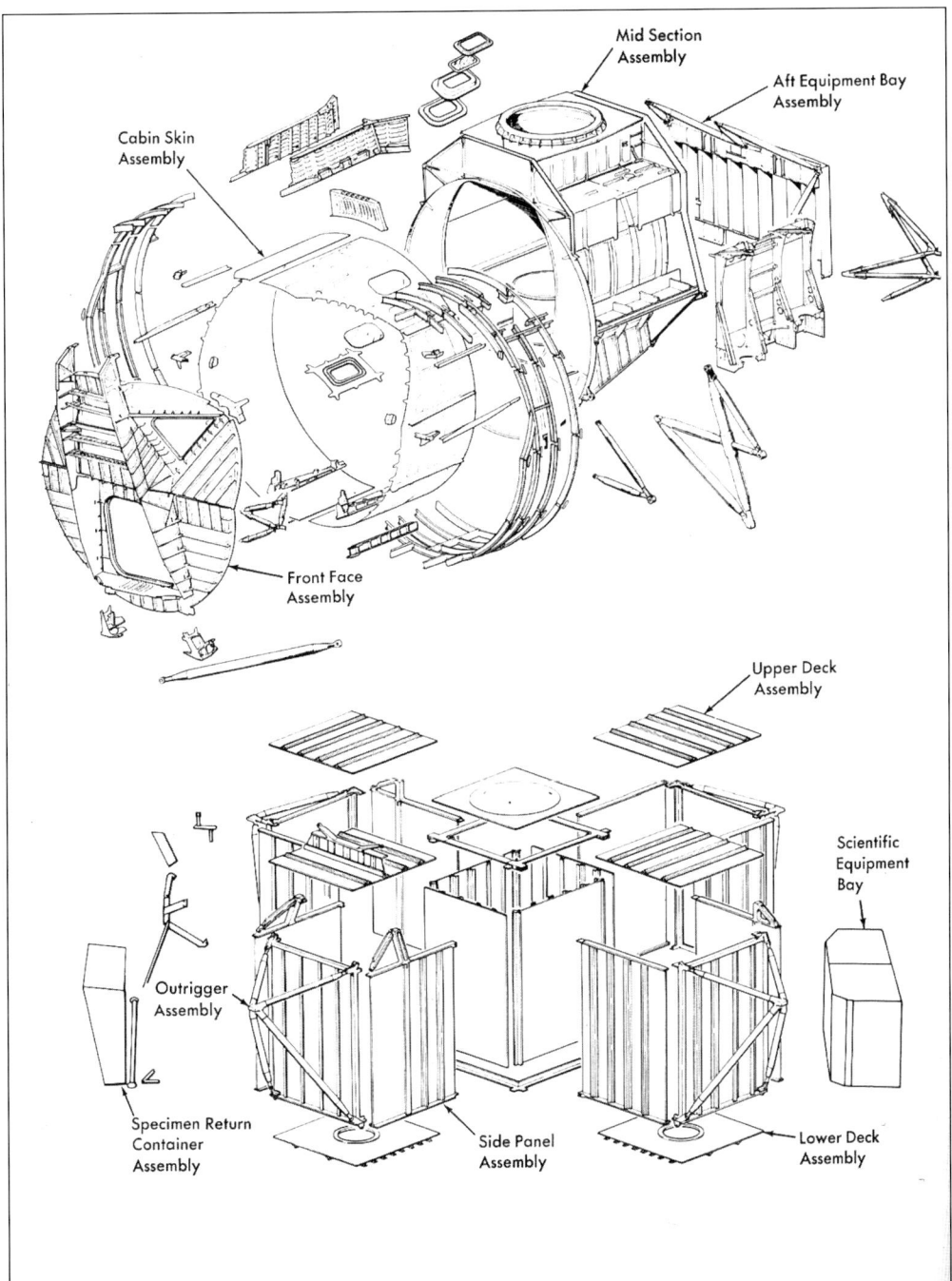

THE BASIC STRUCTURE OF THE LUNAR MODULE. Before the LMs were shipped to the clean room, the basic aluminum and titanium structure was assembled in Grumman's regular factory area. Aluminum skins and stiffeners were milled by hand to make them as light as possible. Every ounce of weight was critical; even bolts were hollowed out. The lighter the vehicle, the more time there was to find a safe landing place.

THE CONSTRUCTION OF THE FRONT FACE OF THE ASCENT STAGE. The front face of an ascent stage nears completion. This was largely machined out of one piece of aluminum with riveted-on stiffeners. This is an early model LM test article with the round hatch. The next step was to attach it to the midsection.

THE CONSTRUCTION OF THE MIDSECTION OF THE ASCENT STAGE. The crew compartment had an overall volume of 235 cubic feet. Structural members were fusion welded wherever possible to minimize cabin air pressurization leaks. This is the beginning of LM-5 *Eagle*, the LM used on the first lunar landing.

THE MIDSECTION REAR OF THE ASCENT STAGE. These early stages of LM construction were done in Plant 3 in Bethpage. Next, the aft equipment bay will be attached to the rear of the ascent stage.

THE COMPLETED ASCENT STAGE. This completed ascent stage test article, PA-1, is now ready to leave Plant 3 for the clean room. The workers who built it pose for one last photograph; their pride in their workmanship is obvious.

THE BASIC CONSTRUCTION OF THE DESCENT STAGE. The cruciform aluminum structure of this descent stage will house the main propulsion system components. The descent engine will be in the center, with four fuel tanks around it. This is LM-2, which was slated to be flown unmanned in Earth orbit. (Photograph courtesy of Northrop Grumman History Center.)

FUEL TANK CONSTRUCTION FOR THE ASCENT STAGE. The fuel for the ascent engine was contained in two identical spherical titanium tanks. The hypergolic components were extremely explosive and detonated on contact, so no igniter was needed, which greatly simplified the system. The fuel was hydrazine, and the oxidizer was nitrogen tetroxide.

THE LANDING RADAR'S ANTENNA. On the bottom of the descent stage was the landing radar that bounced signals off the lunar surface, giving the astronauts an accurate reading of their altitude. The radar's antenna, seen here, was built by Ryan Aeronautical and was protected from the heat of the descent engine by a shield. (Photograph courtesy of Northrop Grumman History Center.)

THE ASCENT STAGE IN THE WORK FIXTURE. The LM stages are now in the clean room and the installation of subassemblies has begun. The electrical and propulsion systems are being installed, and the stages will soon be mated.

PREPARING TO MATE THE ASCENT STAGE. The stage's fuel tanks can now be seen clearly before the cage covering it with shielding is installed. The stage is being raised into position for mating.

MATING THE ASCENT AND DESCENT STAGES. The two stages will soon be bolted together. These bolts will be replaced with explosive bolts at Cape Kennedy. The fuel tanks are covered with protective padding during the construction process. Once mated, the vehicles' buildup will continue.

THE INSTALLATION OF THE LUNAR MODULE IN THE FULL WORK FIXTURE. The two stages have now been mated, and the complete vehicle is being installed in the work fixture, which will fully envelop it. The birdcage structure of aluminum tubing has been installed over the midsection of the ascent stage. The vehicles' thermal and micrometeoroid shielding will be attached soon.

The Plant 5 Clean Room, Looking East, c. 1969. This is one of the most historic buildings in America's aerospace history—it is where every moonship was built. The clean room was specially constructed for this purpose, and it was kept as clean as a surgical room. Up to three LMs could be worked on at one time in the fixtures on the right side. Their average stay in the fixture was 12 months.

THE INSTALLATION OF THE LUNAR MODULE'S SHIELDING. As equipment was steadily being installed inside the LM, work also progressed on the exterior. The ascent stage has now been covered with its aluminized Mylar thermal blanketing, and the thin skins are being secured over it. The descent stage's landing gear outriggers have also been covered with shielding.

LM-5 EAGLE IN ITS WORK FIXTURE. The first spacecraft to land men on the moon is under construction. This is what an LM normally looked like in the clean room. It could hardly be seen.

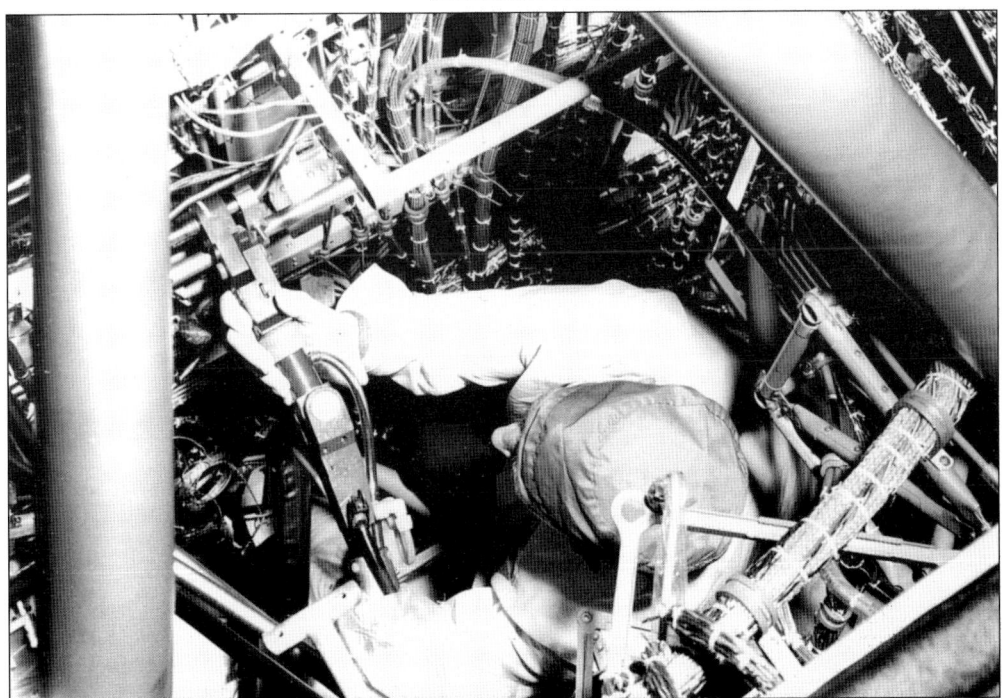

A TECHNICIAN INSIDE THE DESCENT STAGE ENGINE COMPARTMENT. This is probably taking place in the Cold Flow Test Site. The worker is holding a fusion welder to repair a leaking joint in the RCS fuel system that was discovered during testing. The cramped working conditions inside the LM are evident.

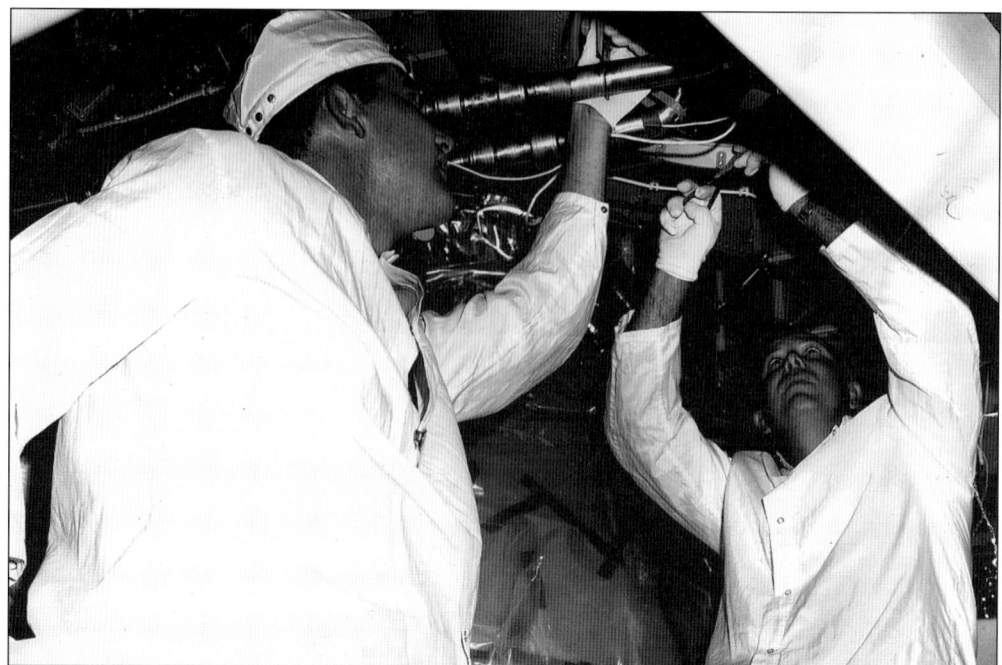

INSPECTING THE FUEL LINE OF THE LUNAR MODULE. Here, beneath a descent stage fuel tank, a Grumman technician and a NASA quality control inspector are examining the fuel lines. Literally every part and every joint on the LM was inspected and signed off on before acceptance by NASA.

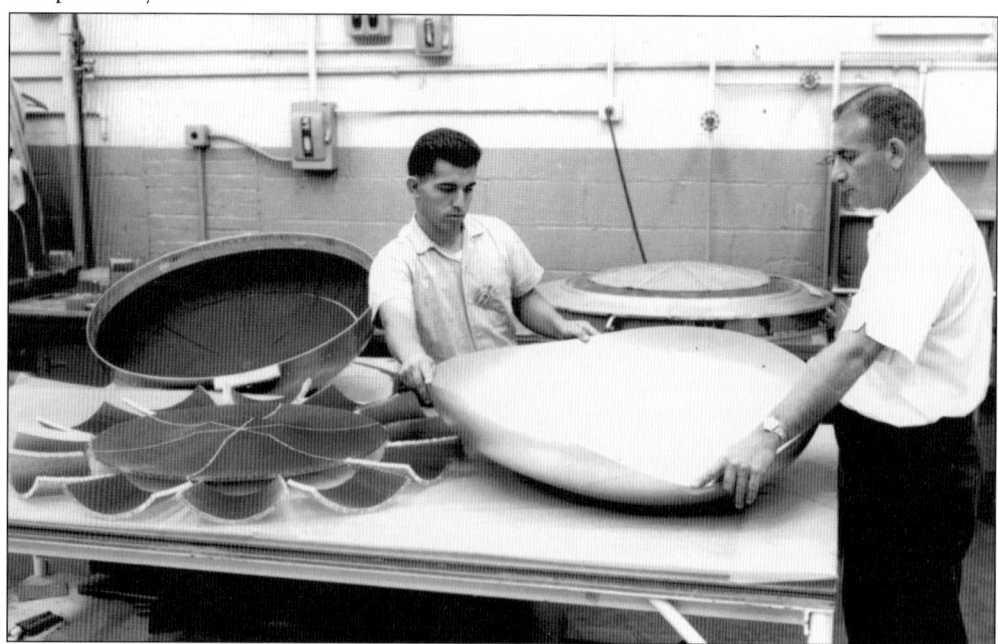

THE CONSTRUCTION OF THE LUNAR MODULE'S FOOTPADS. Each of the LM's footpads was made up of sections of honeycomb aluminum bonded to inner and outer sheets of formed aluminum. The footpads were extremely strong and light. (Photograph Courtesy of Northrop Grumman History Center.)

THE LUNAR MODULE STAGES, SEPARATED FOR TESTING. At several points in the construction process, the stages were sent to the Cold Flow Test Site. There, the fuel lines and valves were tested with nitrogen and glycol was put in the cooling system. The ascent stage's birdcage structure can be seen clearly.

MOVING THE ASCENT STAGE FOR TESTING. Seen here is the ascent stage on its way to the Cold Flow Test Site; its 25-layer-thick thermal blanket is evident. The rendezvous radar in the front is covered with temporary protective padding.

THE LUNAR MODULE ASCENT STAGE AFTER TESTING. Back in the clean room, the LM's aluminum shielding now covers its ascent stage. The large black hose is blowing air in to keep the inside of the vehicle under positive pressure. As the inside is under higher pressure than the outside, this would keep any dirt out of the interior.

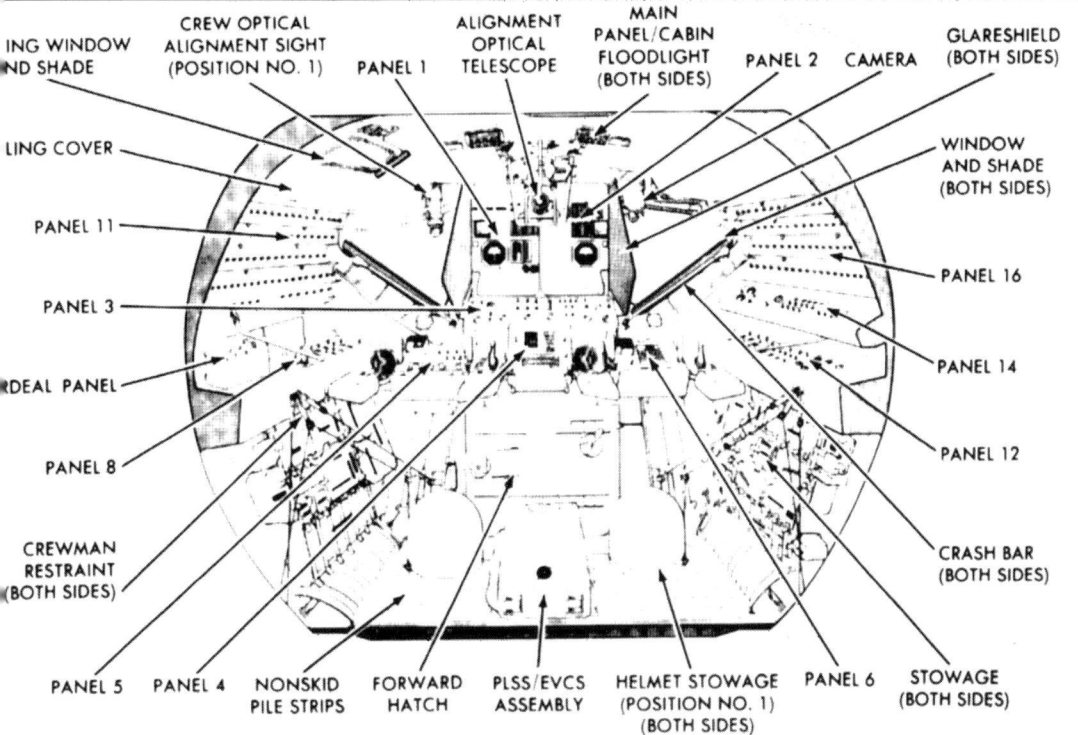

THE LUNAR MODULE'S COCKPIT, LOOKING FORWARD. The commander stood on the left and the LM pilot stood on the right, each looking out his respective window. The controls and displays enabled the astronauts to monitor and manage the LM and control it during separation, docking, and landing.

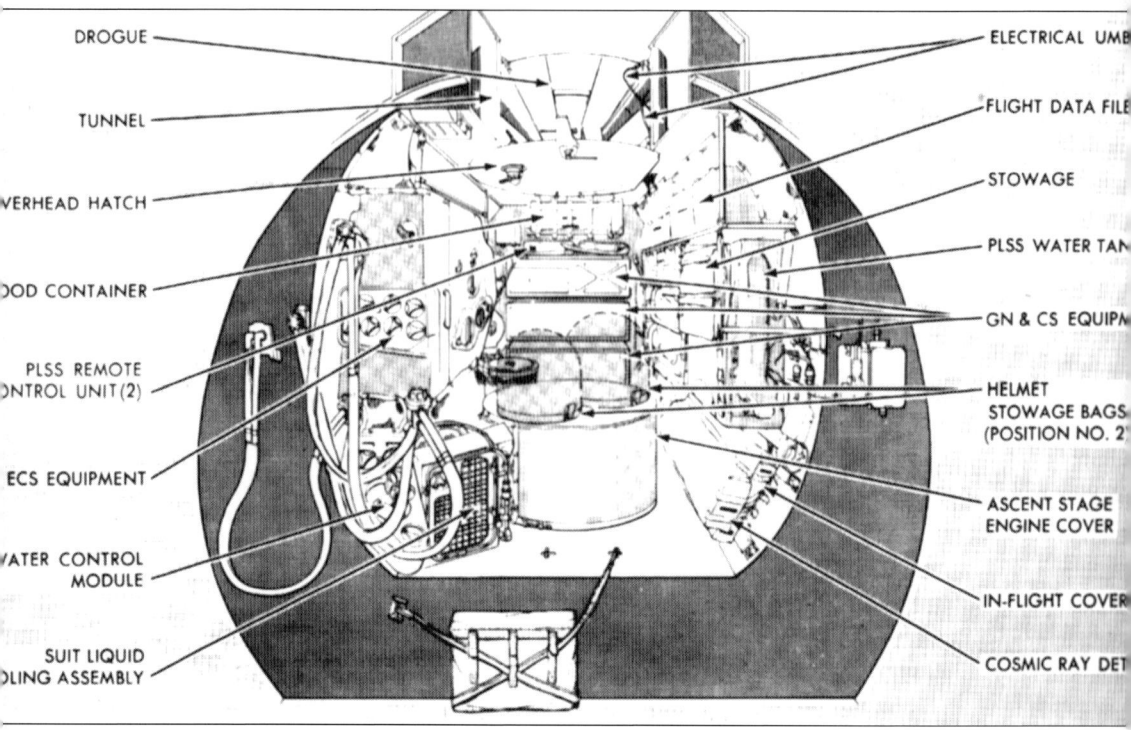

THE LUNAR MODULE'S CREW CABIN LAYOUT, LOOKING AFT. The rear of the cabin contained storage, electrical, and life-support equipment. The ascent engine actually protruded into the cabin and was covered by a cylindrical cover.

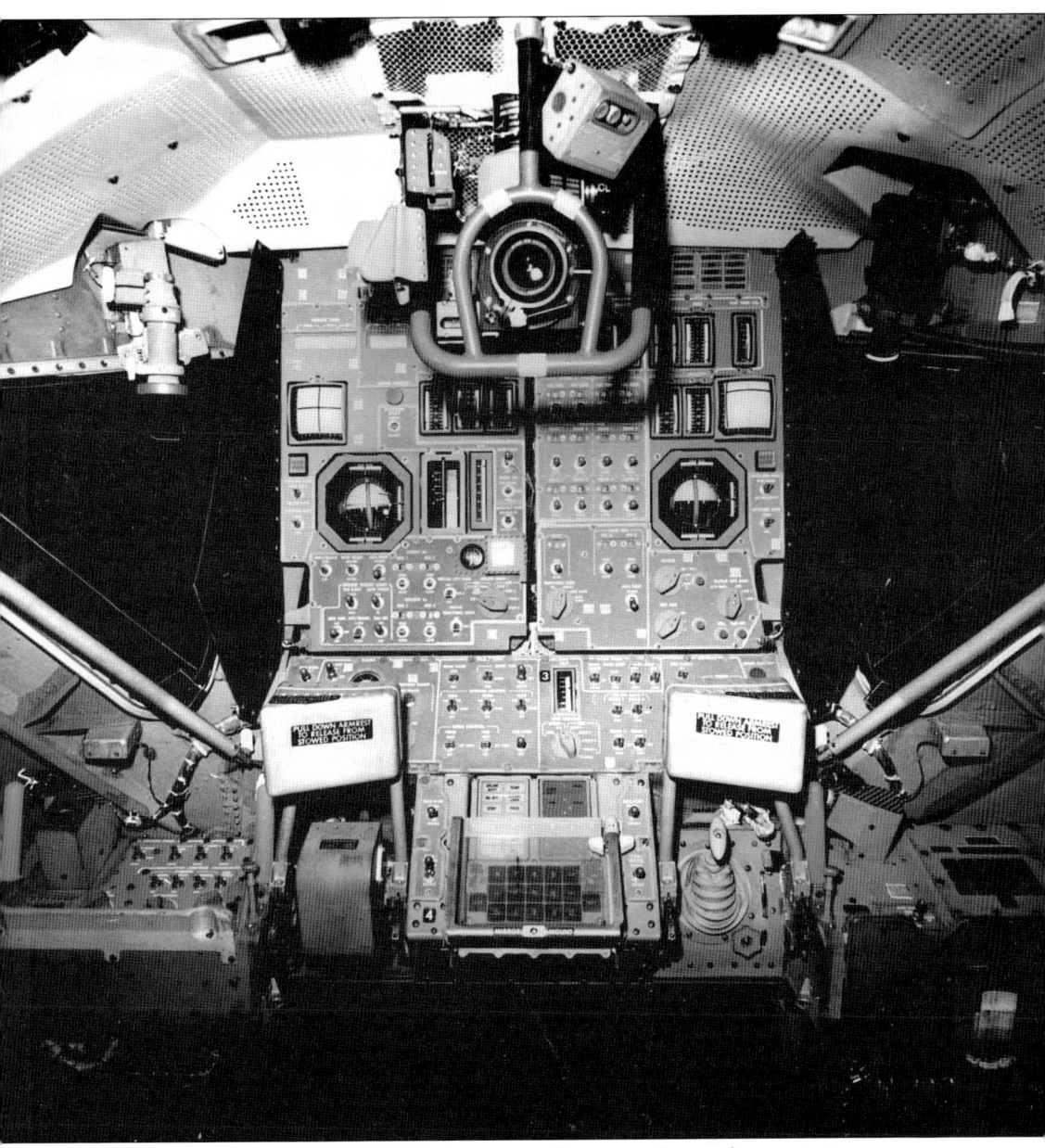

THE LUNAR MODULE'S COCKPIT CONTROLS AND DISPLAYS. Located to optimize astronaut safety and mission success, the primary navigation and guidance readouts, flight computer keyboard, propulsion control, reaction control, environmental control, and flight control and stabilization panels were either shared or duplicated at both stations. In the upper center is the alignment optical telescope, used for navigation and docking. In the lower center is the primary guidance computer keyboard.

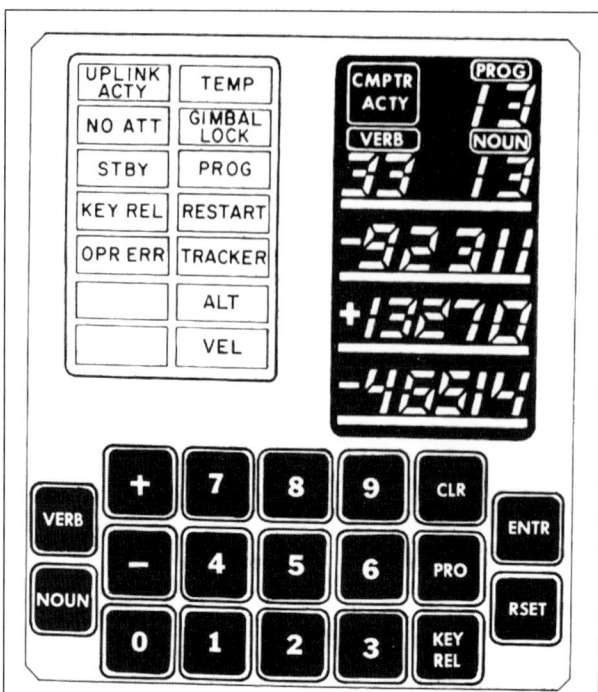

THE LUNAR MODULE'S COMPUTER. The LM's display and keyboard (DSKY) enabled the astronauts to insert data into the primary guidance computer and initiate operations. The astronauts communicated with the computer by pressing a sequence of buttons. "Verb" indicated to the computer that it had to take some sort of action. "Noun" told the computer the next two numerical entries were what the action was being applied to. The 2K memory LM computer is a direct ancestor of all modern PCs, although it had far less memory than a modern pocket calculator.

THE LUNAR MODULE'S DOCKING WINDOW. After Grumman eliminated the front docking hatch, the LM had to use the top hatch to dock with the CM. Thus, the designers provided the commander with a small window directly above his head, through which he could look straight up to align the LM with the CM's docking hatch. The grid on the window indicated distance to the target.

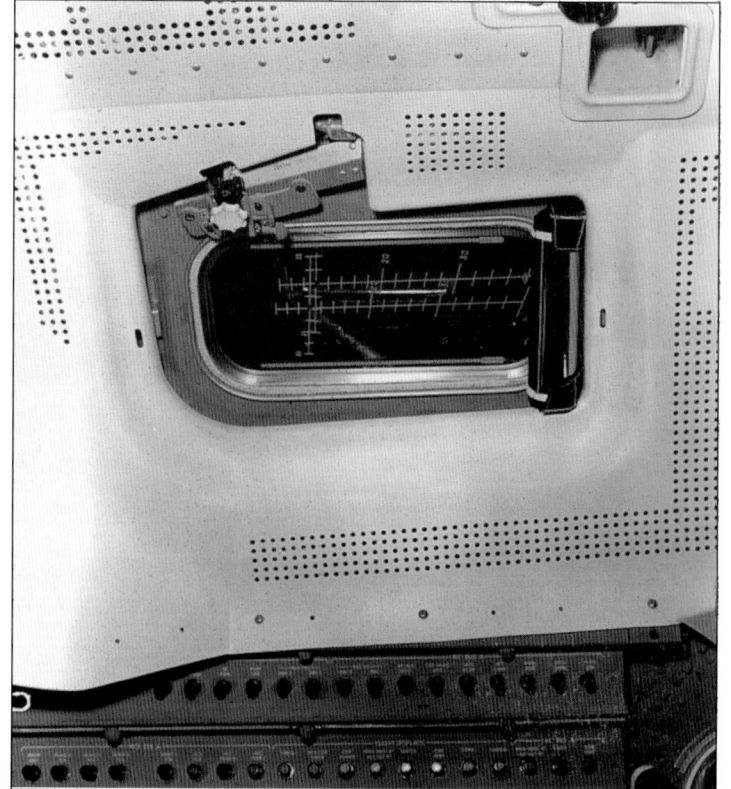

THE ENVIRONMENTAL CONTROL SYSTEM IN THE MIDSECTION. The environmental control system provided a temperature- and pressure-controlled oxygen atmosphere in the LM cabin and crew suits, as well as a supply of water and oxygen for the spacesuit backpacks. It also provided temperature control for the on-board electronic equipment, as well as drinking water. It was built by Hamilton Standard.

THE MIDSECTION, LOOKING UP. The docking tunnel, located at the top rear of the crew compartment, provided the interface between the LM and the CM. It was 32 inches in diameter and 16 inches long.

THE COMMANDER'S INSTRUMENT PANEL. Located on the left side of the central console, this panel contained warning lights, digital counters, navigational instruments, engine thrust control switches, and engine, fuel, and altitude indicators. The black abort button in the lower right would have automatically separated the stages and fired the ascent engine if the LM ran out of fuel during the landing phase.

THE COMMANDER'S STATION. The commander stood directly in front of this window, his left hand on the descent engine thrust controller, his right hand on the joystick that controls the LM's attitude thrusters. The panel in front of him contains the start button for the ascent rocket engine.

THE MIDSECTION, LOOKING DOWN. Shown is the floor of the crew cabin, which has strips of Velcro that adhere to the bottom of the astronauts' boots. In front of the crew, beneath the instrument panel, is the square forward hatch, which is used for egress to the lunar surface. The 32-inch-square hatch is hinged to swing inward. Above it, folded up, are two armrests.

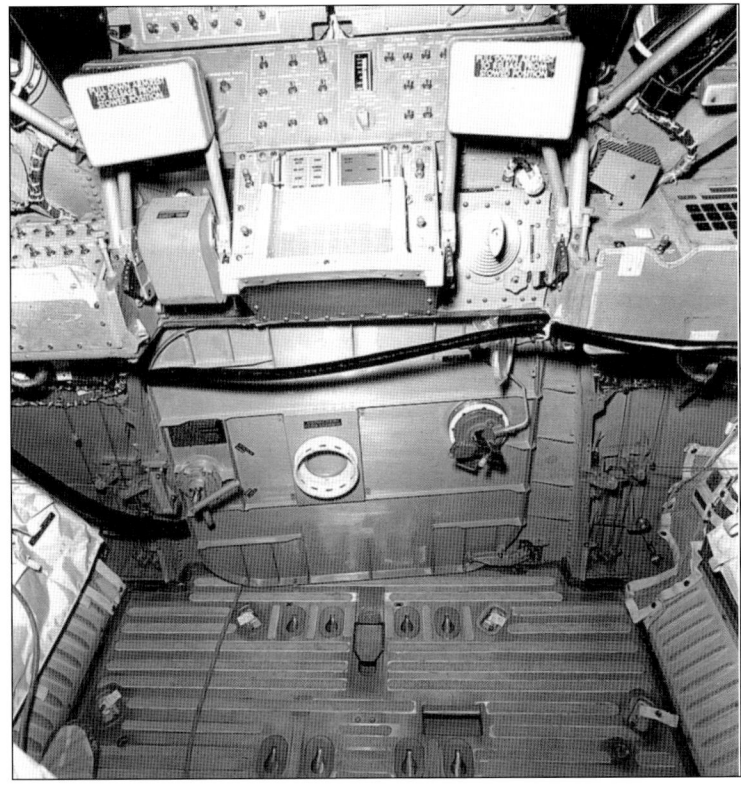

THE RIGHT REAR OF THE MIDSECTION. In the center, one of the astronauts' backpacks (portable life-support system) is stowed, with a pair of lunar overshoes tucked beneath it. Farther back are the rock boxes for the lunar samples.

THE LUNAR MODULE PILOT'S SIDE CONSOLES. Located on the right side of the crew compartment, the center panels contain the controls and displays for electrical power. Above them are the circuit breaker panels.

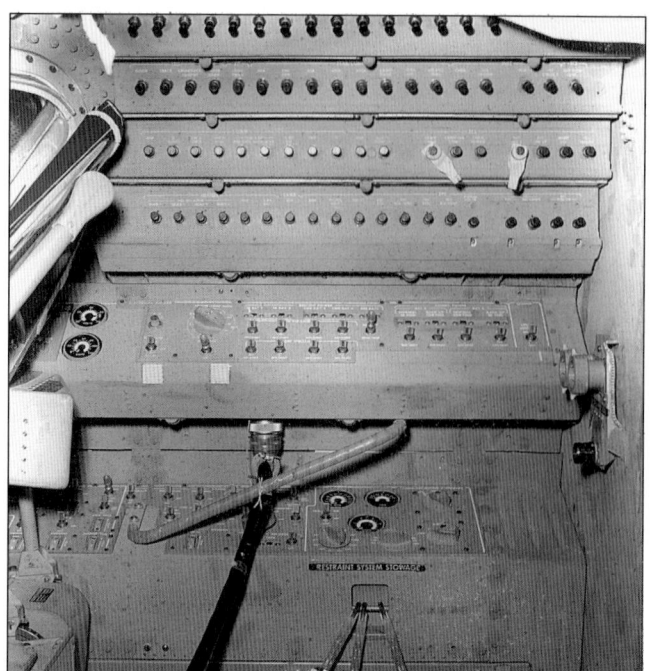

THE DESCENT STAGE BEING MOVED. This descent stage is being temporarily moved out of the clean room, probably on its way to the Cold Flow Test Site or to the rotate-and-clean fixture. On the left is another descent rocket engine ready for installation.

A Completed Descent Stage. This stage is now virtually complete and ready to ship. All of the thermal blanketing and shielding has been attached. The ladder on the front leg was so light that it could only be used in the gravity of the moon, which is one-sixth that of Earth; it would have buckled under a person's weight on Earth. From left to right are Paul Sach (shop supervisor), Ross Fleisig (LM-5 spacecraft manager), and Jay Klevanowski (quality control supervisor). (Photograph courtesy of Ross Fleisig.)

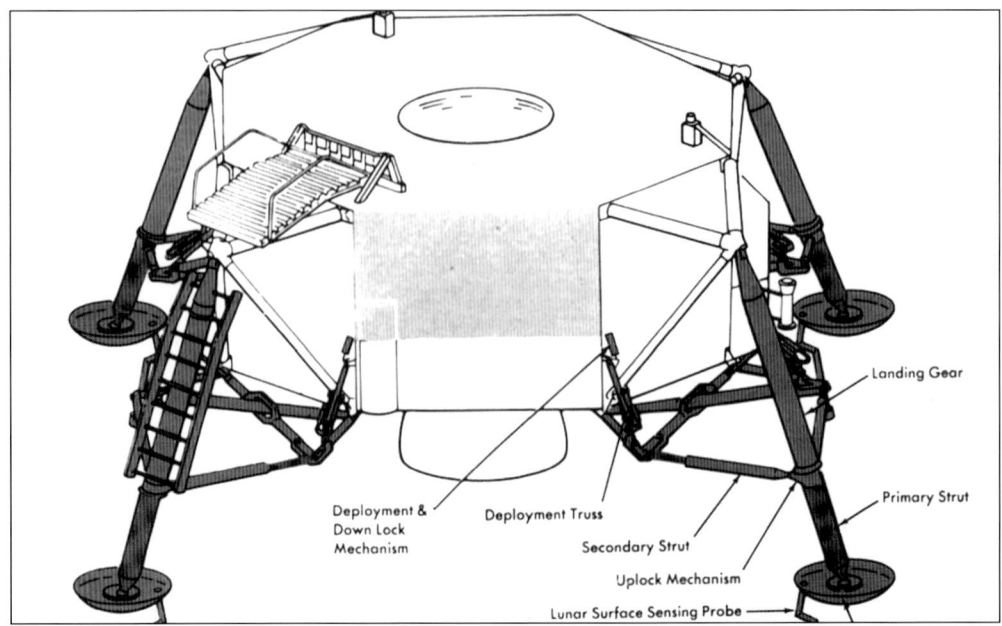

THE DESCENT STAGE'S LANDING GEAR. The cantilevered landing gear folded inward to fit within the Saturn V rocket. The four sets of legs were spring-loaded. They deployed on the way to the moon. The struts were machined aluminum; the footpads were spun aluminum bonded to honeycomb core. Aluminum probes on the footpads were equipped with sensing devices so that the astronauts would get a signal when they were six feet above the lunar surface.

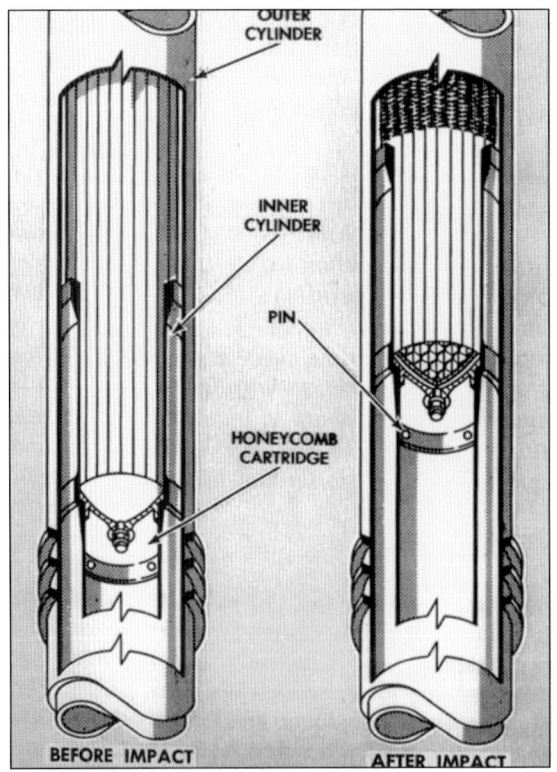

THE LANDING GEAR SHOCK ABSORBER. The LM's landing gear shock absorbers were simple and effective. An inner tube fit within the outer strut and moved upward, crushing an aluminum honeycomb cartridge and absorbing the force of the landing. It only had to work once, and it did.

AN ASCENT STAGE INVERTED IN THE TUMBLER. Several times during the course of the LM's construction, each of its stages was placed in the rotate-and-clean fixture, known as the tumbler. It was literally turned upside down and shaken out for debris that could cause electrical circuits to short or be hazardous to the astronauts in zero gravity.

A DESCENT STAGE IN THE TUMBLER. Here, a descent stage partially through construction takes its turn in the rotate-and-clean fixture. Still to be attached are all of its shielding and the descent rocket engine.

63

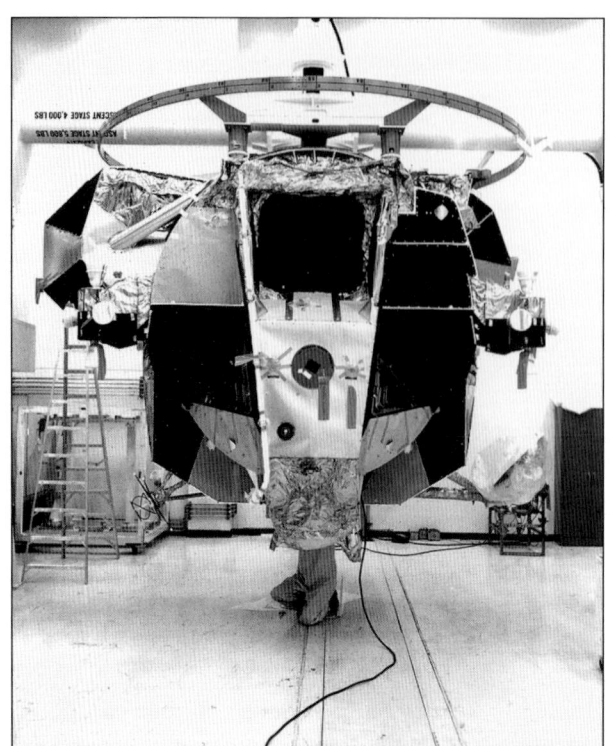

THE COMPLETED ASCENT STAGE IN THE TUMBLER. Ready for shipping, this ascent stage is being given one last turn in the tumbler. The stage is complete with all of its thermal blankets and shielding. The top hatch, where everything fell out, was not given its final installation until after tumbling was complete. This is LM-5, the first LM to land on the moon.

THE DESCENT STAGE IN THE TUMBLER. At this point the stage is virtually complete, with all of its shielding in place. It will be reassembled one more time in the clean room, and its landing gear will be test fitted.

AN ASCENT STAGE ROCKET ENGINE. Grumman technicians examine an ascent rocket engine that has just been received from Bell Aerosystems, the manufacturer. Installing the rocket engines was one of the last major procedures undertaken in the clean room. (Photograph Courtesy of Northrop Grumman History Center.)

REMATING THE LUNAR MODULE STAGES. After returning from the Cold Flow Test Site, the nearly complete LM stages were remated in the clean room one last time. The final systems can now be tested and the landing gear fitted.

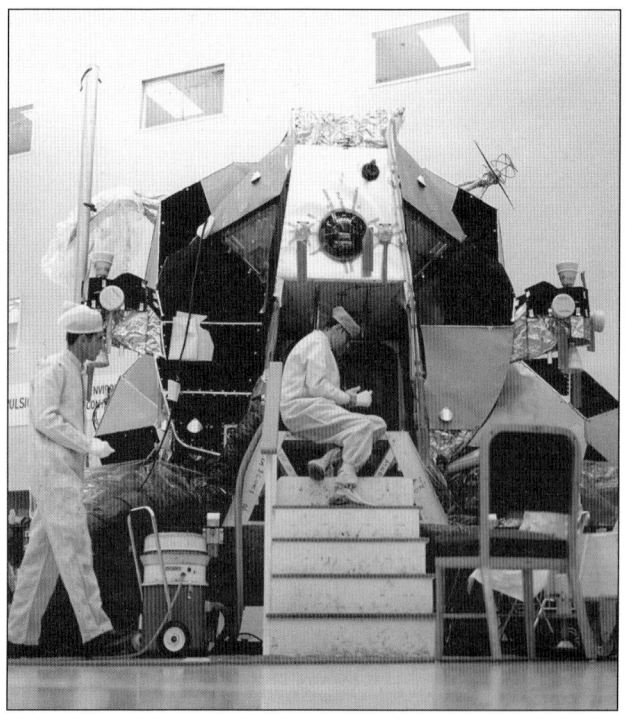

PREPARING TO SHIP THE ASCENT STAGE. The ascent stage is now virtually complete, except for the rendezvous radar that will be finally attached at Cape Kennedy, in Florida. In this photograph, technicians are preparing the LM for shipment. The red tags denote covers that must be removed before launch. The large black hose is still providing positive airflow into the vehicle.

The Ascent Stage Leaving the Clean Room. Now complete, this LM is beginning its trip to Florida. It will soon be sealed inside the plastic covering hanging above it.

A Biological Sampling of the Lunar Module. Prior to leaving Grumman, the LM is sampled for bacteria by the U.S. Public Health Service. Believe it or not, scientists evoked concern that the vehicle might carry to the moon bacteria against which lunar life forms would have no defenses.

Three
Testing and Training for the Moon

LM-5 in the Integrated Test Stand. While the LM is in the clean room, lengthy tests are conducted on every system in the vehicle. Each LM had its own managers and supervisors, whose names and photographs were hanging on the side of the work fixture (lower right). Thus, everyone knew who they were and who was ultimately responsible.

LTA-8 IN THE TEST STAND. This is a top view of the test stand, showing the hose pressurizing the ascent stage through the top hatch. LTA-8 was shipped to Houston, where it was the first and only LM tested in the vacuum chamber at the Manned Spacecraft Center.

A SIDE VIEW OF LTA-8 IN THE TEST STAND. LTA-8 is undergoing an electrical test prior to shipping. Seen here are the technicians. The engineers monitoring the test are about 300 feet away in the ACE room.

ONE OF THE ACE ROOMS IN PLANT 5. The LM-5 team monitors the vehicle during a test. In the ACE room, every system and every valve is checked to ensure smooth operation in flight. The chief test conductor (lower right) is speaking to the technician in the vehicle cabin, who is testing the various components. Entire missions were run through in simulation here.

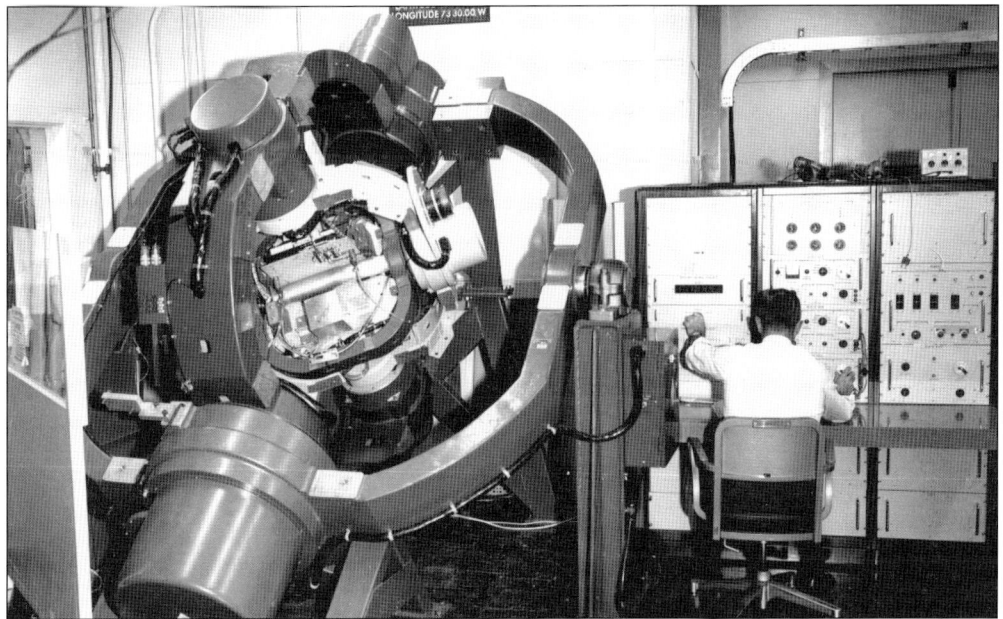

THE THREE-AXIS FLIGHT SIMULATOR IN THE FLIGHT CONTROL INTEGRATION LAB. The guidance system of the LM is being tested here. The primary guidance system was placed in the simulator and was rotated to see how the system responded and how long it would take the gyroscopes to stabilize the vehicle in the event that it went out of control.

THE LUNAR MODULE IN THE COLD FLOW TEST SITE, PLANT 5. Here, the propulsion and cooling systems of the LM were pressurized and checked for leaks. If a leak was found, and they usually were, it was fixed here. (Photos courtesy of Northrop Grumman History Center.)

THE LUNAR MODULE BEING REMOVED FROM THE TEST STAND. This nearly complete LM is being removed from the integrated test stand. It will now be prepared for shipping.

An LM Test Article, Ready for a Drop Test. This LM will be dropped from varying heights to see how the landing gear and shock absorbers are functioning. As these tests were very stressful to the vehicle, only test articles, not flight vehicles, were used.

A LANDING GEAR DEPLOYMENT TEST. This was one of many tests that every LM went through upon its arrival at Cape Kennedy. The vehicle was mounted in an elevated fixture, and the spring-activated landing gear was deployed. (Photograph Courtesy of Northrop Grumman History Center.)

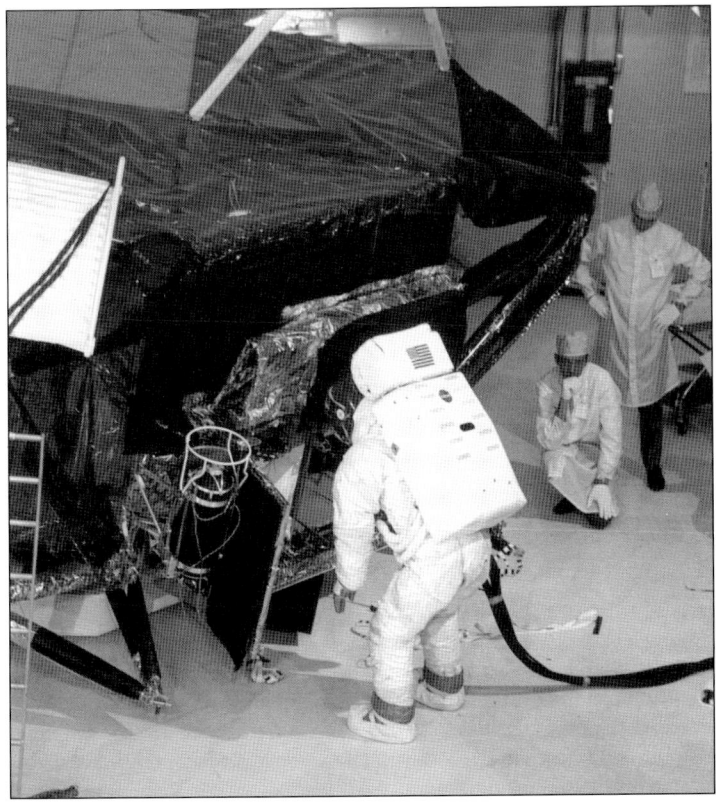

ASTRONAUT TRAINING AT GRUMMAN. The astronauts were an integral part of LM design and testing. Here, an astronaut checks out access to one of the equipment bays of the LM descent stage in the event that the vehicle landed at a severe angle.

NEIL ARMSTRONG IN THE LUNAR MODULE SIMULATOR. Along with LM flight and test articles, Grumman produced two LM simulators in conjunction with the Link company. The LM simulator was identical to the LM crew cabin on the inside, with the addition of projectors and screens in front of the windows to simulate lunar landings. Some crews even practiced sleeping in it. This LM simulator is at the Kennedy Space Center.

A LUNAR LANDING TEST VEHICLE. As it was impossible to fly an LM on earth, NASA developed this device to simulate the final phase of a lunar landing in actual flight. It was powered by a large jet engine in the center with small gas reaction-control rockets on the sides. The astronauts found it very difficult and unstable to fly, and it was ultimately not used by the later crews.

GRUMMAN'S WHITE SANDS TEST FACILITY, 1966. In order to test the LM's propulsion systems prior to actual flight, Grumman established a test site with 340 people at the government's White Sands Test Facility in New Mexico. Test firings using flight-weight test articles duplicated every phase of the lunar mission. Problems discovered here allowed the incorporation of improvements prior to the actual launches. Here, ascent stage test article PA-1 is being lowered into the firing chamber. (Photograph Courtesy of Northrop Grumman History Center.)

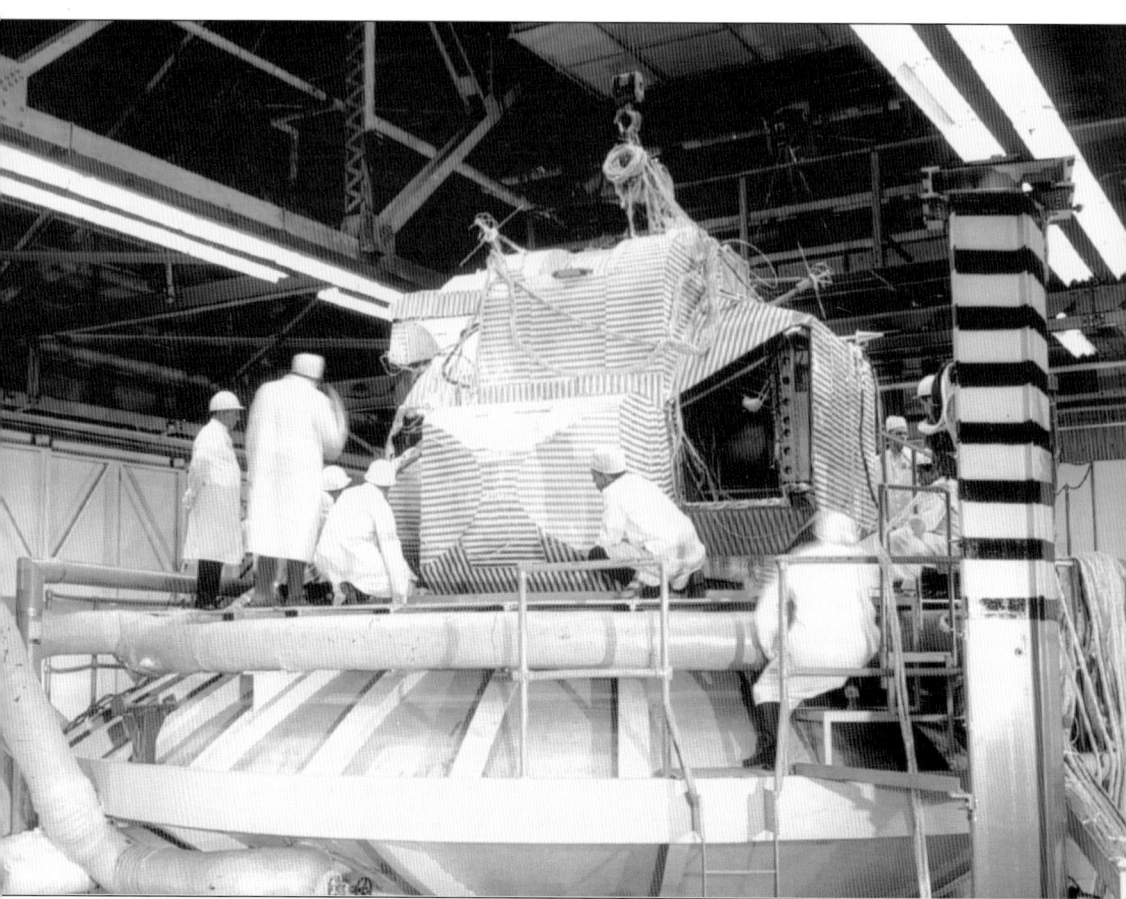

LTA-8 Being Lowered into the Vacuum Chamber, May 1968. LM test article LTA-8 was subjected to three 12-hour tests in Chamber B at NASA's Manned Spacecraft Center in Houston. The tests exposed the LM's systems to the extreme temperatures of space and also simulated the vacuum conditions in Earth orbit. These tests were very successful and cleared the LM for manned flight in Earth orbit. (Photograph courtesy of Northrop Grumman History Center.)

Four
Shipping and Installation

LM-6 Engineers and Technicians Posing in the Clean Room. Before each completed LM was shipped to Florida, the teams that had labored for so long to build it posed for one last photograph. It was the last time they would lay their hands on something they had built that was now on its way to the moon.

LM-9 ENGINEERS AND TECHNICIANS POSING IN THE CLEAN ROOM. All of Grumman's LM workers took great pride in their work on Project Apollo. They knew that this was one of the great moments in American history and that they were an integral part of it.

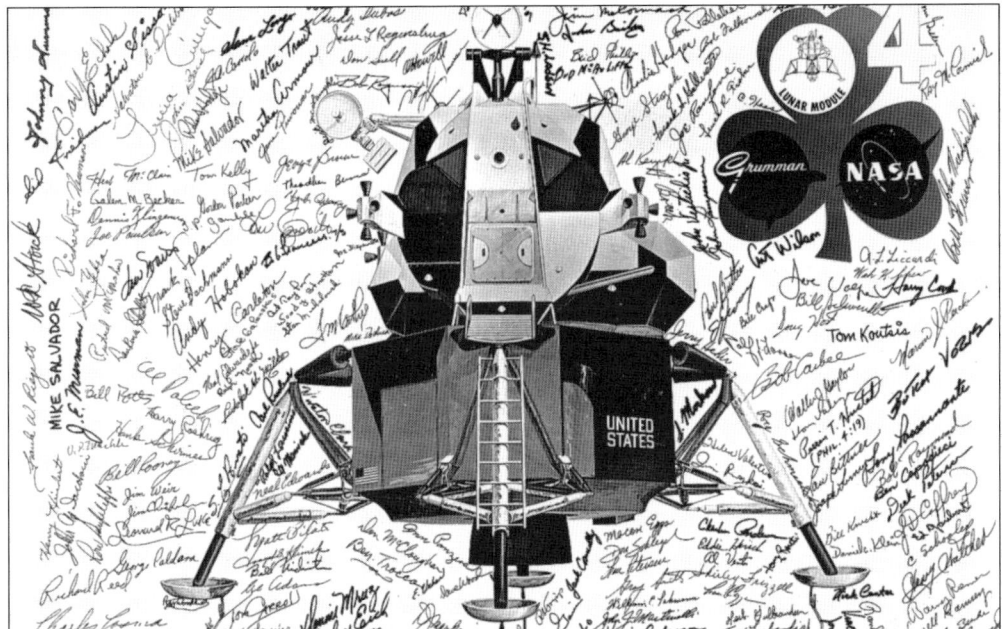

THE LM-4 SIGNATURE POSTER. As the LM was an entirely handmade machine, the workers felt that they had become a part of it. Before each LM was shipped, every worker who had built it signed a poster, which was photographed, reduced, and attached to the side of the descent stage between the layers of thermal blanket. Those signatures remain on the moon to this day.

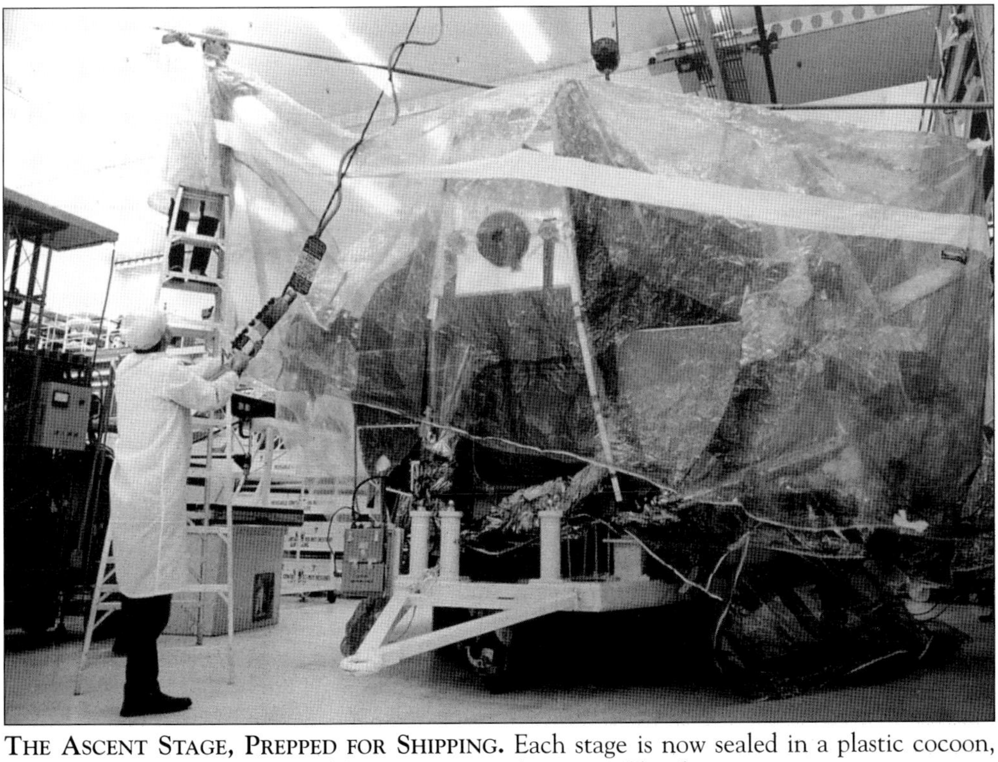

THE ASCENT STAGE, PREPPED FOR SHIPPING. Each stage is now sealed in a plastic cocoon, which will have controlled air blown into it on the trip to Florida.

THE SUPER GUPPY TRANSPORT ARRIVING AT GRUMMAN. The *Super Guppy* was a highly modified Boeing C-97 Stratocruiser with a greatly enlarged upper deck and a hinged nose. It was used to transport the *Apollo* spacecraft, as well as the third stage of the Saturn V. Here, it arrives at Grumman to pick up an LM for the trip south.

LOADING A LUNAR MODULE INTO THE *SUPER GUPPY*. Each stage was sealed into its shipping crate; the LM is now beginning its journey to the moon. A Grumman employee stayed with the LM during the voyage to Cape Kennedy.

GEORGE SKURLA, LUNAR MODULE PROGRAM MANAGER, KENNEDY SPACE CENTER. Once the LMs arrived at Cape Kennedy, it was George Skurla's job to put them through a rigorous testing program and to repair any deficiencies that were found. As the schedule was very tight, the pressure was enormous.

THE LUNAR MODULE STAGES REMATED AT THE KENNEDY SPACE CENTER. Upon arrival, the LM was reassembled and its landing gear attached. Every system was then tested yet again before NASA would give the vehicle its final acceptance.

83

THE LUNAR MODULE IN THE TEST FIXTURE AT CAPE KENNEDY. In this fixture, the landing gear could be tested for proper deployment. Note the large hose still putting positive pressure into the vehicle.

THE LUNAR MODULE MATING TEST. One of the many important tests that every LM went through was the mating test with the CM. In this test, a CM test article was inverted and was mated to the top hatch of the LM. This checked for correct alignment and construction of the docking tunnel and for proper engagement of the latches. (Photograph Courtesy of Northrop Grumman History Center.)

LOWERING THE LUNAR MODULE INTO THE SPACECRAFT LUNAR MODULE ADAPTER. Inside the Vehicle Assembly Building, the LM is gently lowered into the SLA, which then mates to the third stage of the Saturn V. Note the landing legs now in the folded position so that the LM fits.

A TOP VIEW OF THE LUNAR MODULE IN THE SPACECRAFT LUNAR MODULE ADAPTER. The LM is attached to the lower section with four fittings at the ends of its landing gear outrigger struts. Explosive bolts held straps that kept the LM in place until it was released on the way to the moon.

LOWERING THE UPPER SECTION OF THE SPACECRAFT LUNAR MODULE ADAPTER INTO PLACE. Once the LM was secured, the upper section of the SLA was lowered over it. The upper section had four folding panels that opened to expose the LM once it was successfully on its way to the moon.

LIFTING THE COMPLETED APOLLO SPACECRAFT INTO POSITION. Once the LM was secured and covered in the SLA, the *Apollo* command and service modules were lowered into place on top of it.

PLACING THE APOLLO SPACECRAFT ON TOP OF THE SATURN V ROCKET. The complete LM and CM assembly is placed on top of the S-IV stage of the rocket. The complete vehicle stood 363 feet tall. Even once the LM was inside the rocket, workers had access to the LM via a catwalk hundreds of feet in the air. They continued to run power and leak tests right up until the vehicle was fueled for launch.

Five
MISSIONS TO THE MOON

APOLLO 9, LM-3, IN EARTH ORBIT. The first manned test flight of the LM, on *Apollo 9*, LM-3, only operated in Earth orbit. Here, the four panels of the SLA have been jettisoned, and the CM will now dock with the LM while still attached to the S-IV stage.

LM-3 in Earth Orbit, March 1969. Every system on the LM was tested on this mission, and each performed well. Here, the legs have been extended. Note the probes extending from the bottom of each footpad. These had sensors on them that told the astronauts when they were six feet off the moon so they could shut down the descent engine. The probe on the front leg was deleted on all subsequent LMs for fear it would bend up upon landing, making the ladder unusable.

THE LM-3 ASCENT STAGE PRIOR TO DOCKING. The descent stage has now been jettisoned, and the ascent engine has been test fired. They will shortly redock and transfer back to the CM. The LM was left in Earth orbit and burned up on reentry; it was never meant to come back to Earth.

A PLUME DEFLECTOR MOCK-UP. The biggest problem found on *Apollo 9* was that the exhaust from the downward-firing RCS thruster impinged on the descent stage, scorching it. To prevent this from happening in the future, plume deflectors were fitted below this thruster to deflect the exhaust away. Grumman's Saul Ferdman (left) explains the concept to NASA's Wernher Von Braun.

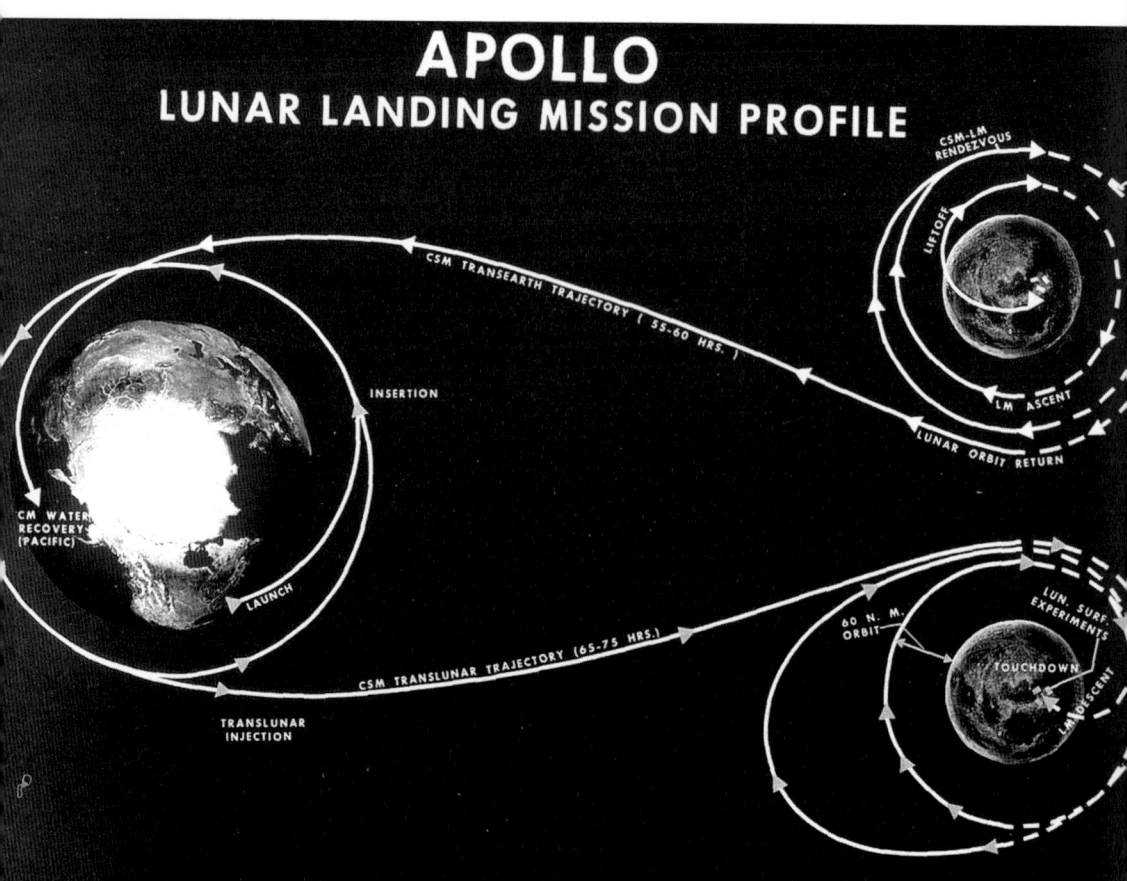

THE LUNAR MISSION PROFILE. The *Apollo*'s mission was to take off from the surface of the earth, which was traveling 1,000 miles per hour as it rotated; go into orbit at 18,000 miles per hour; speed up at the proper time to 25,000 miles per hour; travel to a body in space 240,000 miles distant, which was itself traveling 2,000 miles per hour relative to Earth; orbit around this body; and drop a lander to its surface—and to do this with near pinpoint accuracy and perfect timing.

APOLLO 10, LM-4, ASCENT STAGE IN LUNAR ORBIT, MAY 1969. The *Apollo 10* mission was the first and only rehearsal in the lunar environment. The LM descended to within nine miles of the moon, the stages separated, and the ascent stage redocked with the CM in lunar orbit.

THE CREW OF APOLLO 11. Pictured from left to right are Neil Armstrong, commander; Michael Collins, CM pilot; and Edwin "Buzz" Aldrin, LM pilot. Armstrong and Aldrin were the first human beings to set foot on a world that was not their own.

THE LAUNCH OF APOLLO 11, JULY 16, 1969. LM-5 *Eagle* leaves the earth for the moon. The mighty Saturn V rocket was the most powerful machine yet built by man.

TRANSPOSITION AND DOCKING. While on its way to the moon, the CM separated from the SLA, rotated 180 degrees, docked with the LM, and extracted it from the S-IV stage. Once free, the landing legs could be extended.

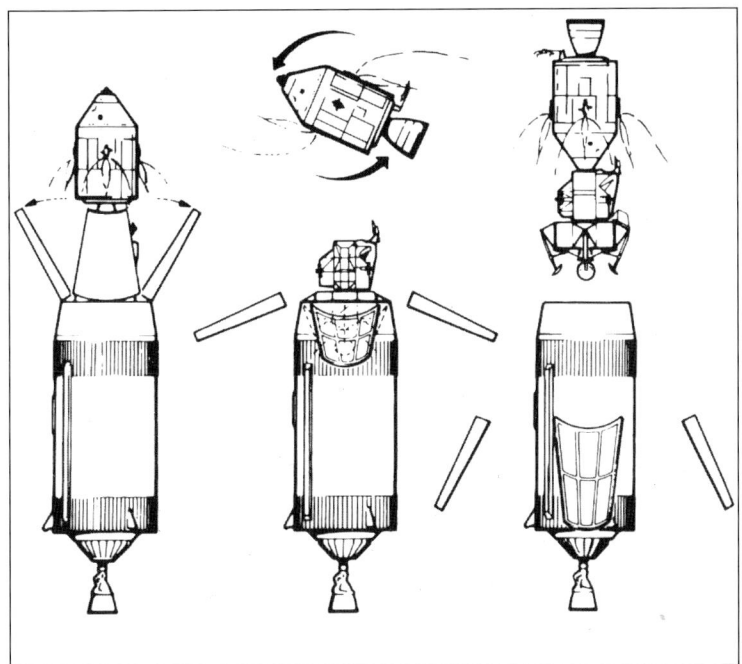

LM-5 EAGLE SEPARATING FROM THE CM. Now in lunar orbit, Neil Armstrong and Buzz Aldrin have undocked from the CM and are ready to descend to the surface. The vehicle first rotated 360 degrees so that Michael Collins, in the CM, could make a visual inspection.

THE LUNAR MODULE DESCENDING TO THE MOON. At this point the LM is about 60 miles up. Armstrong and Aldrin are harnessed in the cockpit in a standing position. At precisely the right moment, the descent engine will be fired for the 12-minute landing phase.

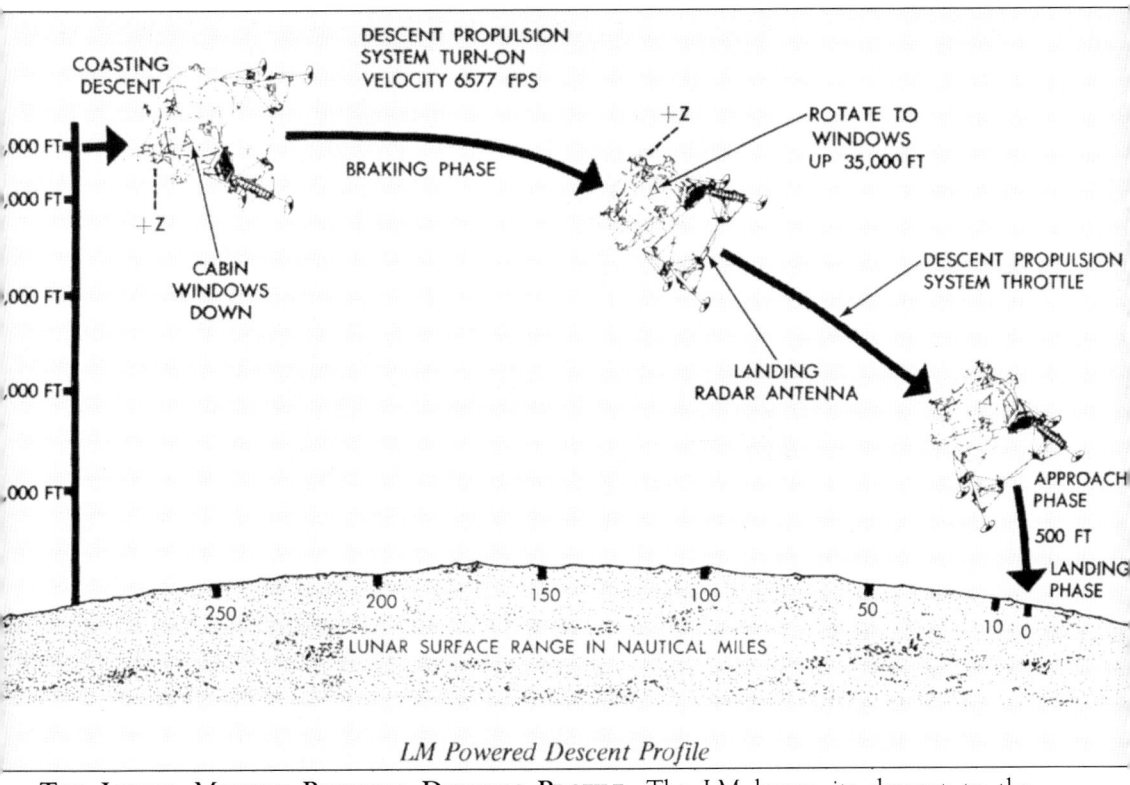

LM Powered Descent Profile

THE LUNAR MODULE POWERED DESCENT PROFILE. The LM began its descent to the moon upside down and backwards. At 35,000 feet it rotated so that its windows were facing upwards; this position allowed the landing radar to begin picking up the lunar surface. Only in the last 10,000 feet was the LM pitched forward so that the astronauts could pick out a clear landing site.

THE LUNAR MODULE ENGINEERING ROOM, BETHPAGE, JULY 20, 1969. With all eyes on the television, the engineers who designed the LM watch intently to see how it performs. Like the rest of the world, they do not know how the mission will turn out, but they hope for and expect the best. Note the many rolls of hand-drawn blueprints used in the days before PCs.

PLANT 3 WORKERS, BETHPAGE, JULY 20, 1969. The men who built the structure of the LM closely follow the mission. With their own hands, they built something that would now land on another world.

THE PLANT 5 CLEAN ROOM, JULY 20, 1969. The technicians working on the next LMs in line pause long enough to watch the first lunar landing. The astronauts' lives depended on the quality of these technicians' work.

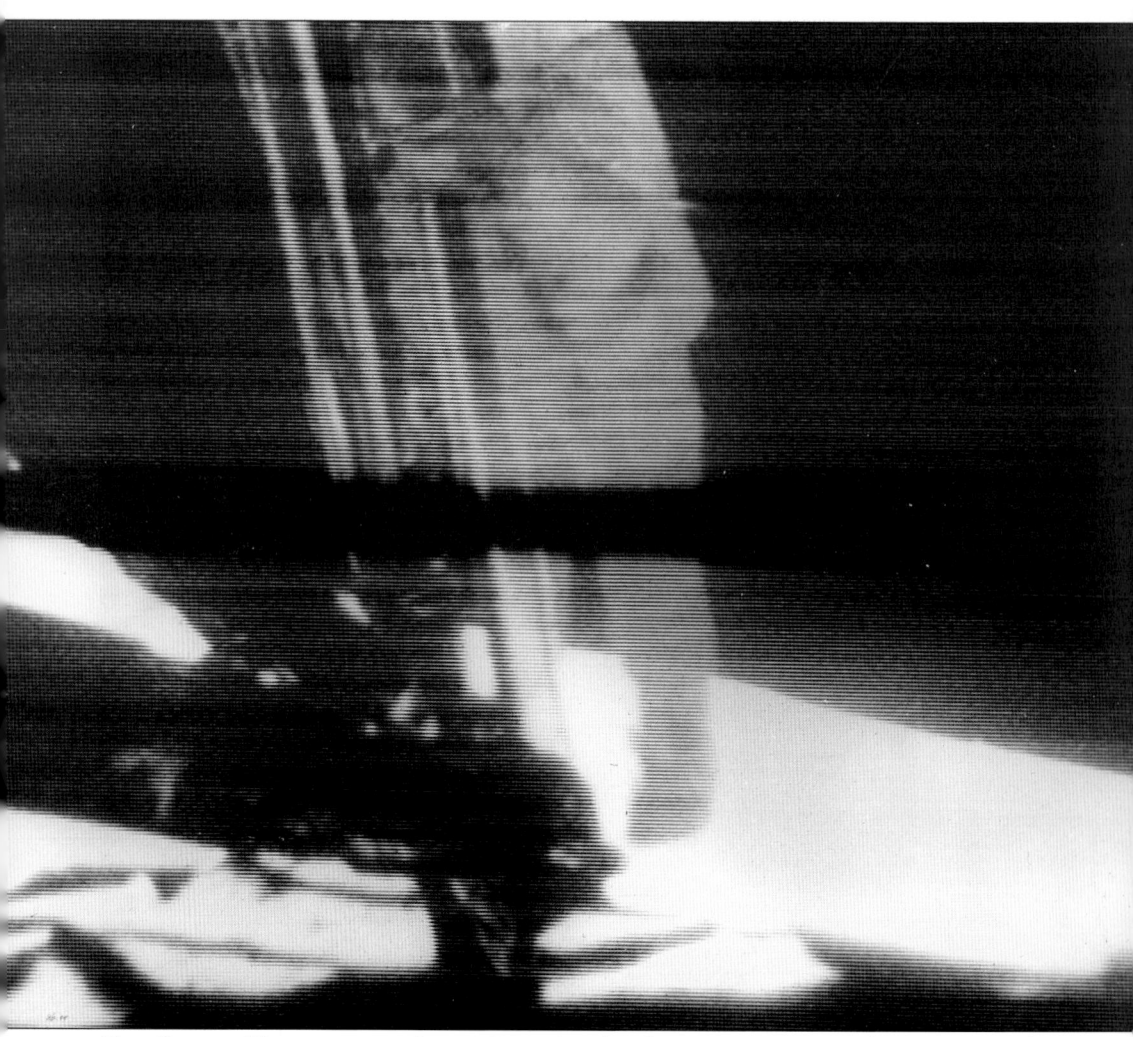

THE SEA OF TRANQUILITY, JULY 20, 1969. A television camera on the LM descent stage records Neil Armstrong about to make "one small step." In spite of concerns of deep lunar dust, the LM sank in only about an inch. Armstrong also landed the LM more softly than expected, resulting in very little compression in the shock absorbers. This resulted in a good-sized first step from the LM ladder to the lunar surface.

Buzz Aldrin Descending the Lunar Module's Ladder. LM-5's two-and-a-half-year journey is now nearly complete, as it has successfully landed men on the moon. From Long Island to the moon, it has performed flawlessly. To this day its descent stage remains there, in quiet testimony to what is possible by man.

HERE MEN FROM THE PLANET EARTH
FIRST SET FOOT UPON THE MOON
JULY 1969, A. D.
WE CAME IN PEACE FOR ALL MANKIND

NEIL A. ARMSTRONG   MICHAEL COLLINS   EDWIN E. ALDRIN, JR.

The Plaque on the LM-5's Leg. This plaque on the LM's front leg, protected by a thermal blanket during the landing, was unveiled by Armstrong once the LM had landed on the moon. The aluminum plaque was mounted between the third and fourth rungs of the ladder.

LM-5 ON THE SEA OF TRANQUILITY. Buzz Aldrin is now deploying the scientific experiment package. The relatively smooth, flat *Apollo 11* landing site was carefully selected. Later missions landed in more difficult places.

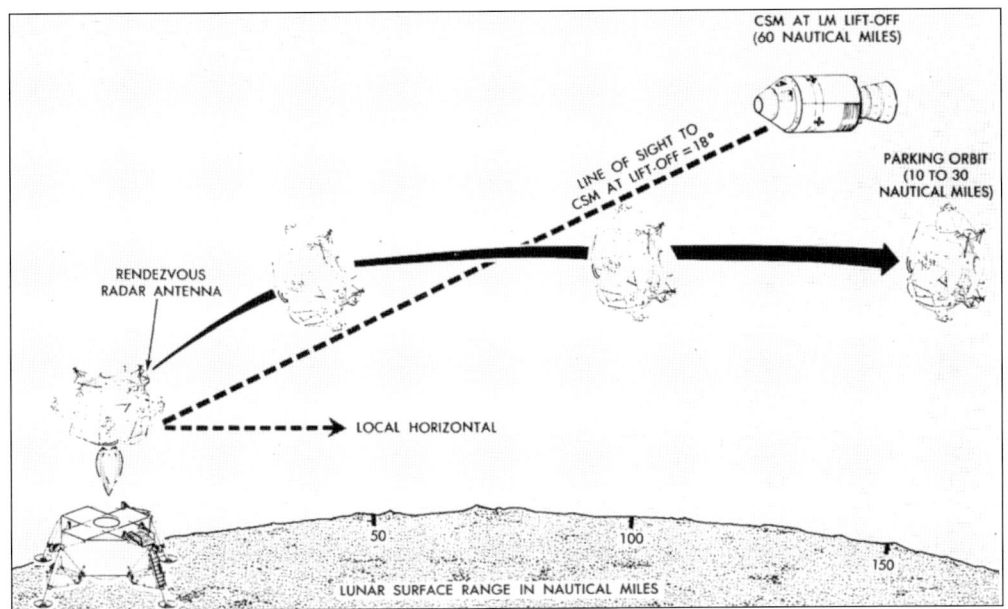

THE LUNAR MODULE'S ASCENT PROFILE. The liftoff from the moon was the most nerve-racking part of the mission for all those who designed and built the LM. Everything had to work perfectly at the moment of liftoff. Four explosive bolts holding the stages together had to blow, a heavy metal blade had to cut through five inches of electrical cabling connecting the stages, and the ascent engine had to fire. It all did, every time.

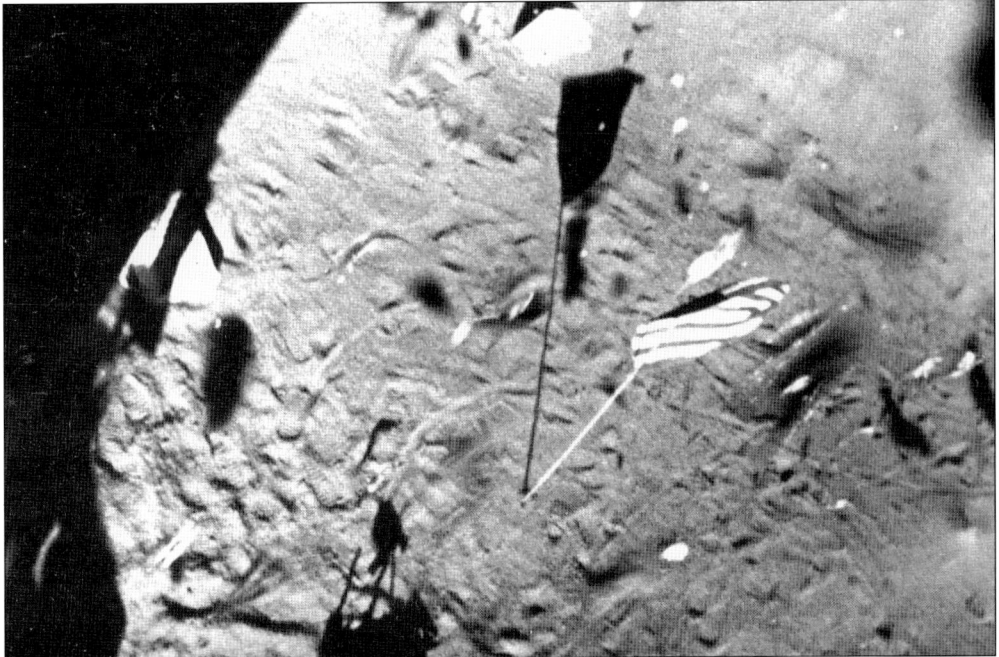

THE VIEW FROM INSIDE THE LUNAR MODULE AT THE MOMENT OF LIFTOFF. The ascent from the moon turned out to be a very smooth ride. However, the astronauts were surprised at the amount of debris blown off from the descent stage. Buzz Aldrin also noted that they planted the flag too close to the LM; the last thing he saw was it being blown over.

The Rendezvous and Docking in Lunar Orbit. The ascent from the moon and the docking with the CM went fairly smoothly. Without the heavy weight of the descent stage, the astronauts thought the LM handled like a sports car.

LM-6, APOLLO 12, ON THE OCEAN OF STORMS, NOVEMBER 1969. Astronauts Pete Conrad and Alan Bean made a pinpoint landing, coming within 600 feet of their intended target, an old Surveyor robot lander.

THE DAMAGE TO THE SERVICE MODULE, APOLLO 13, APRIL 1970. On the way to the moon, an explosion in the service module, just behind the CM, caused the astronauts to abort the mission and fight for their lives. In order for its crew to survive, the LM had to be used as a lifeboat and a tugboat, a role that was never fully planned for.

MODIFICATIONS TO THE LIFE-SUPPORT SYSTEM, APOLLO 13. The LM that was built to keep two men alive on the moon for two days now had to keep three people alive for a week. Instructions were sent up to modify the life-support system to handle the extra carbon dioxide. The LM had to run on minimal power to stretch the batteries; coolant for the electronics was also critical. The astronauts had to use the LM engines to steer and propel them back to Earth and to align them for reentry.

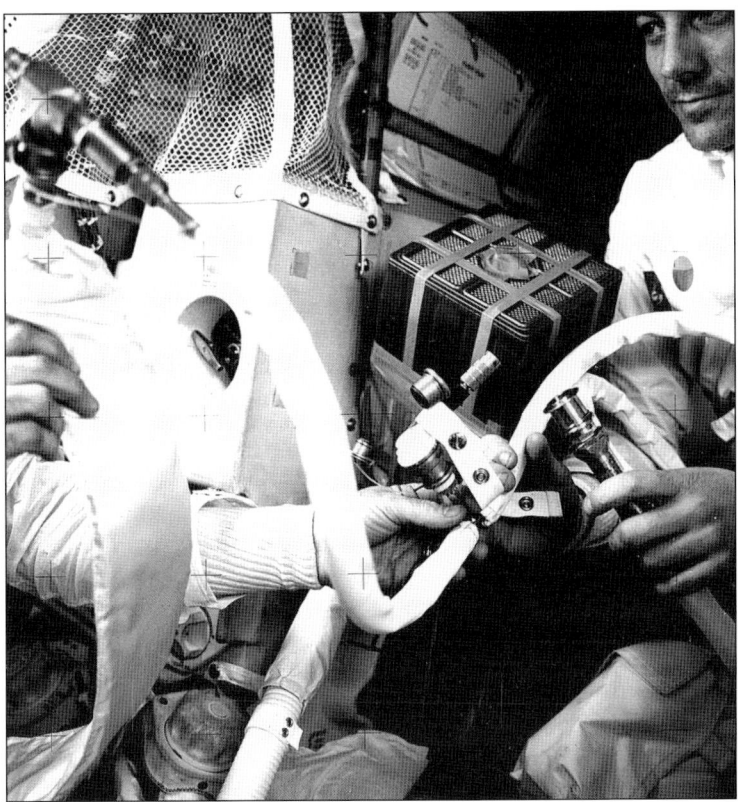

THE APOLLO 13 ASTRONAUTS AT GRUMMAN, MAY 1970. The *Apollo 13* crew made a special trip to Grumman to thank the LM designers and builders for creating such a great spacecraft. It saved their lives in a way no one had ever anticipated.

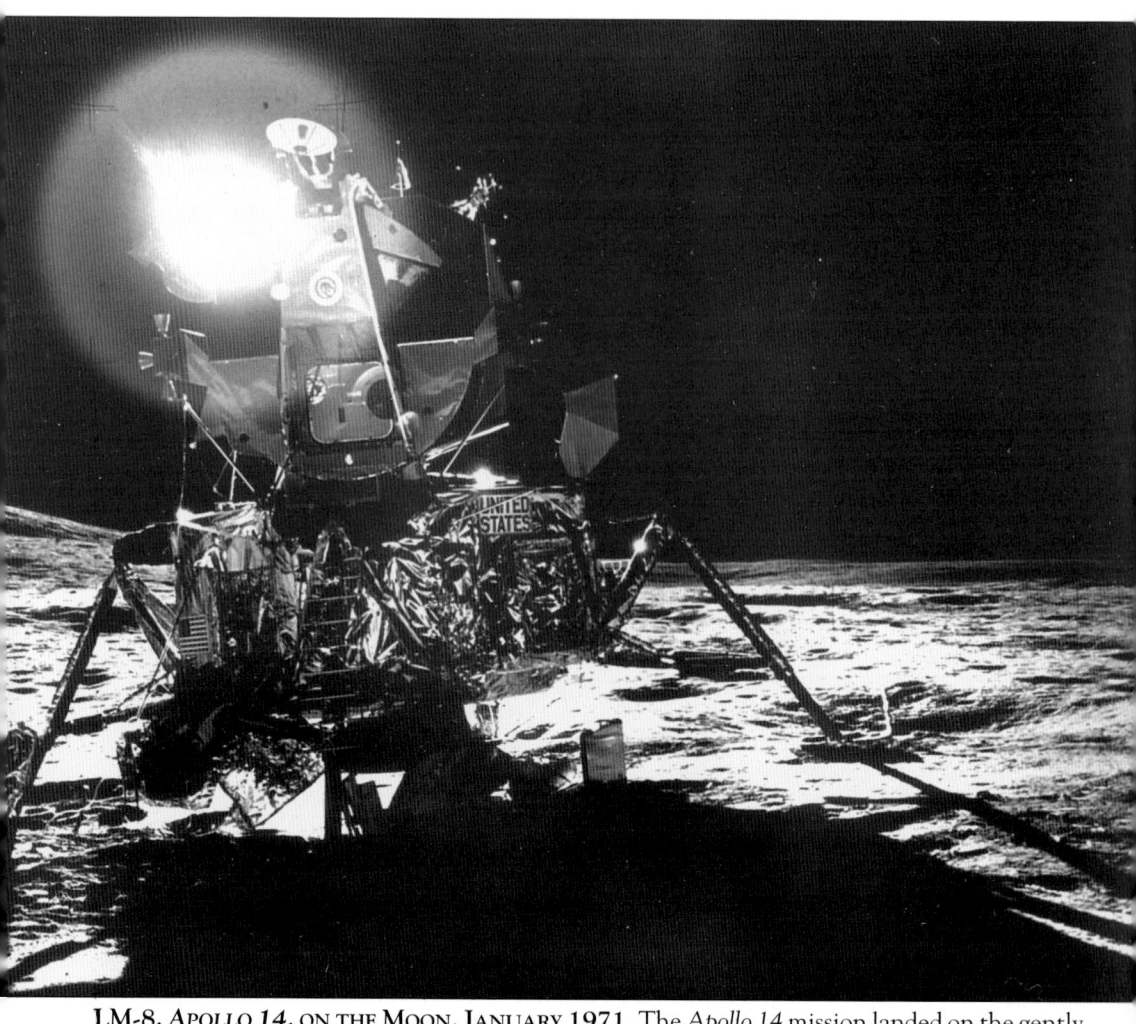

LM-8, Apollo 14, on the Moon, January 1971. The *Apollo 14* mission landed on the gently sloping side of the Fra Mauro crater. This was the first exploration of the lunar highlands.

LM-10, APOLLO 15, ON THE MOON, JULY 1971. This mission landed in the Hadley Apennines mountains. The astronauts lived on the moon and explored for two and a half days. It was the first mission with a lunar roving vehicle.

LM-11, APOLLO 16, ON THE MOON, APRIL 1972. *Apollo 16* landed in the Descartes region of the moon. Cmdr. John Young works at the lunar roving vehicle prior to deployment of the science experiments package.

LM-12, APOLLO 17, ON THE MOON, DECEMBER 1972. *Apollo 17* made a precision approach and landed at the bottom of the Taurus-Littrow Valley. The astronauts lived on and explored the moon for 75 hours. They returned with 243 pounds of lunar samples.

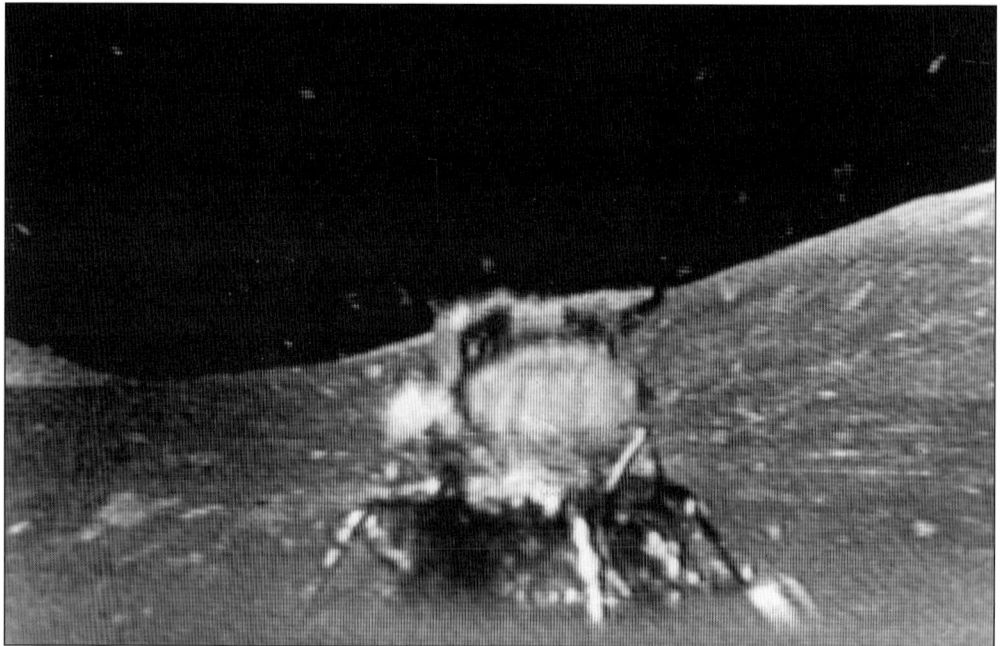

THE LIFTOFF FROM THE MOON, APOLLO 17, DECEMBER 14, 1972. This photograph, taken remotely by the TV camera on the lunar rover, shows the ascent stage as it blasts off to return to orbit. This was the last day that human beings walked on the moon.

Six

THE LUNAR MODULES THAT NEVER WERE

A PROPOSED GRUMMAN LUNAR BASE. Grumman hoped that Project Apollo would continue beyond the first few lunar landings. Thus, the company designed specialized LMs to be used on future missions. Here is a proposed moon base, with three or more LMs that would land in the same area so the astronauts could live and explore for months. In front is the standard LM; in the rear is a descent stage with Molab; on the right is an LM descent stage fitted with living quarters and a laboratory.

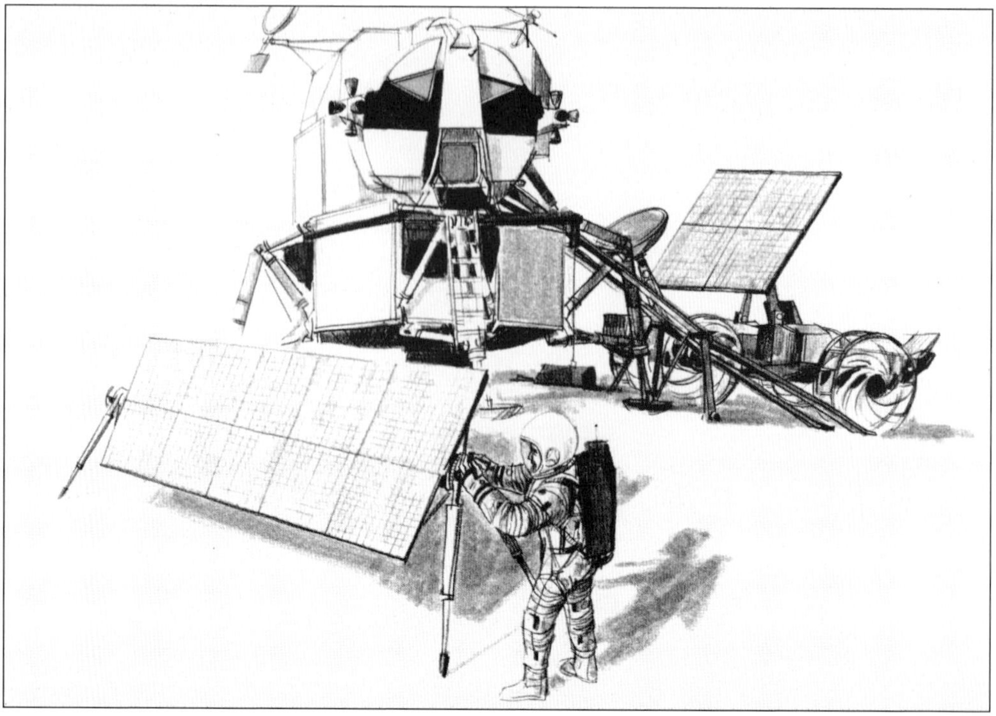

A LUNAR MODULE SHELTER PROPOSAL. The LM shelter was designed to make an unmanned landing on the lunar surface, remain stored for up to three months, and then support two men for a month. No ascent capability was required.

A LUNAR MODULE TRUCK PROPOSAL. The LM truck consisted of a modified ascent stage capable of an unmanned automatic landing from lunar orbit. It was to carry a large payload of 10,000 pounds—living quarters, communications equipment, roving vehicles, scientific experiments, and additional life-support stores—in order to greatly extend lunar exploration. Once the truck had arrived safely, the manned LM was to have landed nearby.

A MOLEM PROPOSAL. In this concept, the entire LM ascent stage was transformed into a long-range, pressurized lunar roving vehicle, designed to greatly expand the astronauts' exploration capability. The vehicle was to land unmanned and drive off a modified descent stage. The manned LM, landing nearby, would be used for the astronauts' return to Earth.

A Stellar Lunar Module Proposal. Once it became clear that there were to be no further lunar landings, Grumman developed ideas to use LMs in Earth orbit. The solar-powered stellar LM was designed to be a free-flying orbital observatory, manned and serviced as needed. It was to have the capability of observing the universe free of the pollution and distortion of the earth's atmosphere.

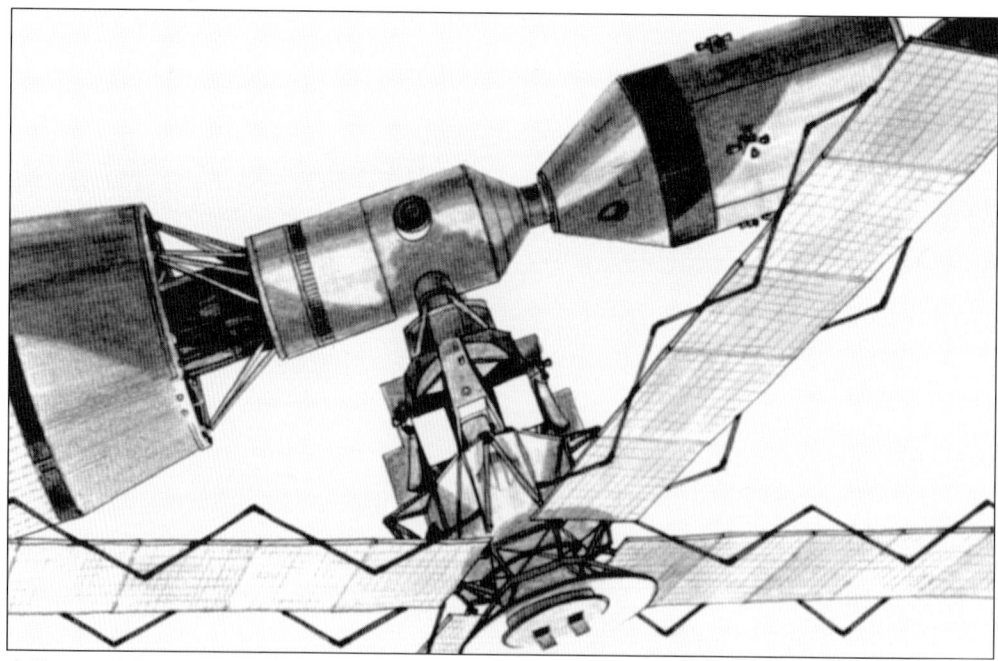

A Lunar Module Apollo Telescope Mount Proposal. For a time there was thought of placing a highly modified LM on the Skylab space station of the mid-1970s. In this LM the descent stage replaced by various astronomy experiments, solar panels, and control gyroscopes. Ultimately, officials decided to equip Skylab with a new small telescope mount for this purpose.

A LUNAR MODULE ARTIFICIAL GRAVITY LABORATORY CONCEPT. Another proposal was to modify and equip an LM with very long cables to attach it to a CM in Earth orbit. The whole assembly was designed to rotate to see if a space station could create artificial gravity in orbit.

A LUNAR MODULE LABORATORY CONCEPT. The LM lab was designed to have one or two LMs that were highly modified and equipped with experiment instrumentation capable of operating in Earth obit or lunar orbit. The lab was to be provisioned to sustain two astronauts for 45 days in space. Stripped of its ascent and descent propulsion systems and the landing gear, the LM lab was still to have retained its docking ability with the CM.

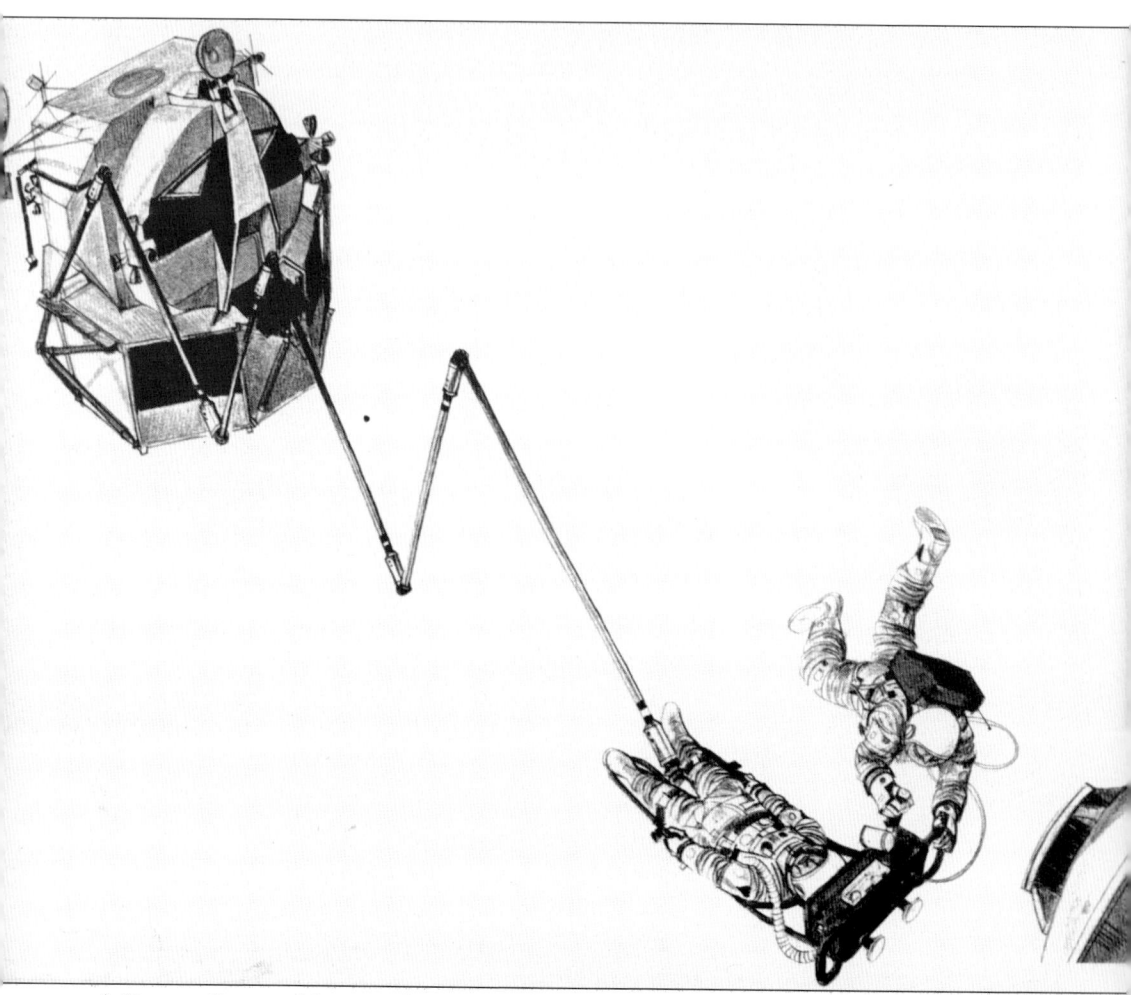

A Rescue Lunar Module Concept. The LM's highly efficient two-stage propulsion system, combined with its reaction control system and versatile guidance and navigation capability, was designed to offer an in-orbit, highly maneuverable manned vehicle of unique capability. It could have intercepted and rendezvoused with other spacecraft or astronauts requiring medical assistance. The Air Force briefly considered using one to disable Soviet satellites.

A NASA SPACE TUG CONCEPT. Hoping that Project Apollo would continue through the 1970s, NASA planned to develop a reusable LM at Grumman. The design consisted of a modified descent stage with a cylindrical ascent stage that could have been reused as a ferry between the earth and the moon. Future lunar landing vehicles will probably resemble this.

LM-2 AT THE NATIONAL AIR AND SPACE MUSEUM, WASHINGTON, D.C. Today, three flight article LMs that were never used remain on Earth. LM-2 was to be flown unmanned in Earth orbit; however, the successful test flight of LM-1 precluded this.

LM-9 AT THE KENNEDY SPACE CENTER, FLORIDA. LM-9 was to have been modified and flown on the *Apollo 18* mission. It now hangs in the Apollo–Saturn V Center.

LM-13 at the Cradle of Aviation Museum, Garden City, New York. LM-13 was to have flown on the *Apollo 19* mission. It is the only LM currently exhibited on a simulated lunar landscape. (Photograph courtesy of Frank Pullo.)

Seven
Lunar Module Vehicle Logos

The Grumman Kennedy Space Center Logo. This patch was worn by Grumman employees assigned to test and install the LM at Cape Kennedy. Along with an *Apollo* mission logo, each lunar module had its own Grumman vehicle logo. Here are all known examples, published for the first time.

THE LTA-8 LOGO. This test article was the first and only LM tested in the vacuum chamber in Houston in 1968. The successful operation of all LM systems in the simulated vacuum of space allowed the LM to be man rated for *Apollo 9*.

THE COLD FLOW TEST SITE LOGO. The Cold Flow Test Site, in Bethpage, conducted the critical tests of the LM's propulsion and cooling systems in order to clear the craft for flight status.

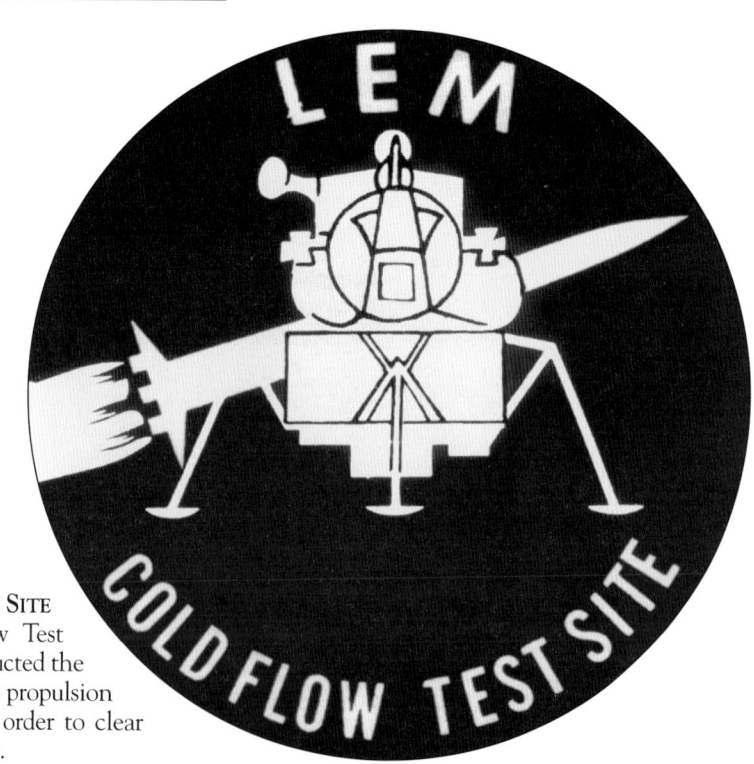

THE LM-1 LOGO. This LM was flown unmanned in Earth orbit on Apollo 5. Successful testing of all systems led to the LM being approved for manned spaceflight.

THE LM-4 LOGO. LM-4, named *Snoopy*, flew on the *Apollo 10* mission and was tested in lunar orbit.

THE LM-5 LOGO. LM-5 was the famed *Eagle* of the *Apollo 11* mission. It was the first to land on the moon.

THE LM-6 LOGO. LM-6, named *Intrepid*, was flown to the moon on *Apollo 12*.

THE LM-7 LOGO. LM-7, named *Aquarius*, saved the astronauts on *Apollo 13*.

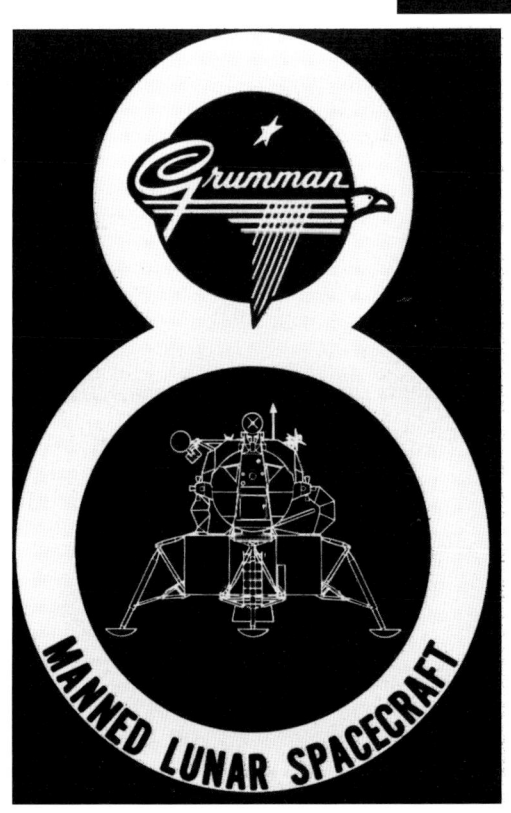

THE LM-8 LOGO. LM-8, named *Antares*, flew to the moon on *Apollo 14*.

THE LM-9 LOGO. LM-9 was never used. It is now on exhibit at the Kennedy Space Center.

THE LM-10 LOGO. LM-10, *Falcon*, was the first extended-stay LM; it flew to the moon on *Apollo 15*, landed, and stayed for two and a half days.

The LM-11 Logo. LM-11, named *Orion*, flew to the moon on *Apollo 16*.

The LM-12 Logo. LM-12, *Challenger*, was the last LM to be used; it landed on the moon with *Apollo 17*.

THE LM-13 LOGO. LM-13 was to fly to the moon on *Apollo 19*, but the mission was canceled. It is now on exhibit at the Cradle of Aviation Museum.

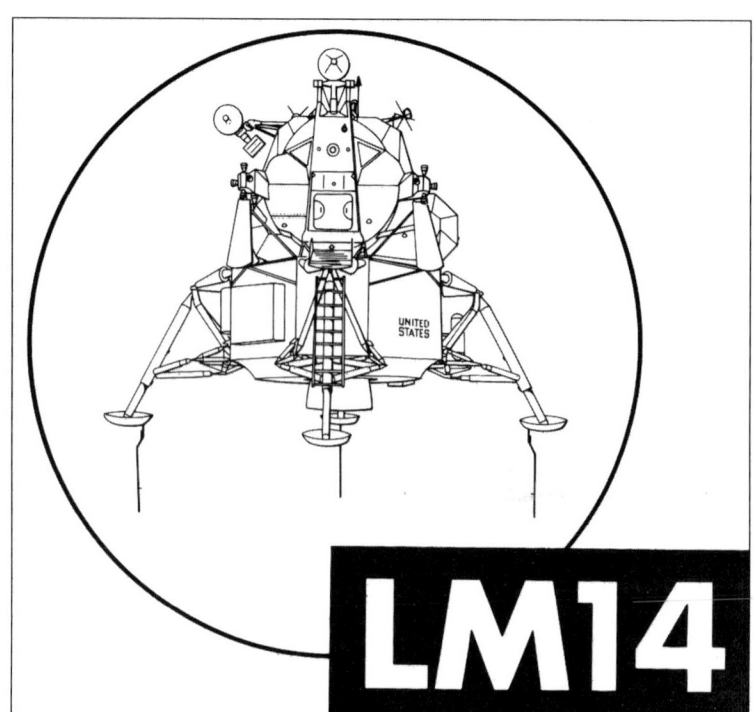

THE LM-14 LOGO. This LM was slated to be used on the *Apollo 20* mission. The mission was canceled, and the vehicle was scrapped.